T0254904

Post Normal Accident

Post Normal Accident

Revisiting Perrow's Classic

Jean-Christophe Le Coze

CRC Press is an imprint of the
Taylor & Francis Group, an informa business

CRC Press
Taylor & Francis Group
6000 Broken Sound Parkway NW, Suite 300
Boca Raton, FL 33487-2742

Printed on acid-free paper

International Standard Book Number-13: 9780367502287 (Paperback)
International Standard Book Number-13: 9780367483999 (Hardback)

Visit the Taylor & Francis Web site at
http://www.taylorandfrancis.com

and the CRC Press Web site at
http://www.crcpress.com

Contents

Introduction ix
Author xv

1. One Book, Two Theses (*NA and na*) 1
Introduction 1
The Reception of *Normal Accident* 2
Singling Out and Comparing High-Risk Systems in the 1980s 2
Critics of Perrow's Technological Determinism 4
Component Accidents versus Normal (or System) Accident 5
Hopkins' Refutation of *Normal Accident* 5
Hopkins' Omission: The Extended Version of Normal Accident 7
Charles Perrow's Sociology 9
Seven Related Books 9
A Set of Interwoven Studies 9
Perrow's Sociology of Organisations 11
The Context of the 1960s 11
Four Key Dimensions: Technology, Structure, Goals and Environment 13
The Society of Organisations 15
A Summary of Perrow's Sociology 17
Normal Accident, the Second Thesis 18
Reintroducing the Book 18
Sources of Tension When Writing *Normal Accident* 19
Beyond the Technological Rationale 20
The Case of the Challenger 21
A Neo-Weberian Approach to Disasters 21
Task: A Cognitive Layer of Analysis 22
Goal and Structure: An Organisational/Managerial Layer of Analysis 23
Environment: A Macrosystem Layer of Analysis 25
Summary 25
Implications of the Second Thesis 28
What Does "Normal" Mean? 28
Does the Book *Normal Accident* Make Any Sense? 29
Was Perrow Right for the Wrong Reasons? 30
A New Version of Normal Accident? 31
Some Limits of Perrow's Argument 31
Summary of Chapter 1 32

2. Hopkins, the Unofficial Theorist of *NA* 35
Introduction 35
Hopkins, the Storyteller 36
Studying and Visualising Accidents 36
What Do We See in This Picture? 36
Visualisations of Hopkins and Perrow 41
Writing Successful Stories 43
A Narrative Structure 43
An Example of Narrative Structure: Auditing 44
Theory of the Second Thesis 49
The 1980s: Critical White-Collar Crime Model of Accident 49
2000s–2010s: A Normative Theory of Safety 51
Technology (Task), Structure, Goal and Environment 54
(Safety) Culture 56
Back to the Longford Case 58
Confronting the Second Thesis of *Normal Accident* 58
The Complexity Argument 58
A Sophisticated, Hidden, Normative Model of Safety 59
A New Formulation of the *High-Reliability Organisation versus Normal Accident* Debate? 61
Summary of Chapter 2 65

3. Errors from the Top 67
Introduction 67
The Obviousness of Strategy 68
Failing Executives and Corporate Malfeasance 68
Organisations at and beyond the Limits 71
What We Know about Human Errors 73
Safety as Strategy 74
The Importance of Strategy for Businesses 74
Linear, Adaptive and Interpretive Views of Strategy 75
Power of Executives and Top Managers 76
Framing Strategy 78
What We Know about Strategic Failures 79
Analysing Strategic Failures 79
Degree of Strategic Failure 81
Defining and Summarising Strategy in the Context of *Post Normal Accident* 82
Illustrating Safety as Strategy 84
Silo Case 86
Pyro Case 87
Petro Case 89
Safety as Strategy 90
Making Sense of the Three Cases from a Strategic Angle 90
Strategic Mistake, Failure or Fiasco? 92

A New Hindsight Bias? 94
Problem of Strategy or of Strategy Implementation? 95
A New Reductionism? 96
Summary of Chapter 3 98

4. From Component to Network Failure Accidents 99
Introduction 99
Globalisation: A Very Short Overview 100
Globalisation, a Central Notion 100
A First Controversy: Is Globalisation Really New? 101
A Second Controversy: Is Globalisation Good or Bad? 102
A Third Controversy: Is Globalisation a Process of Uniformity? 102
Beyond the Controversies 103
Connecting Globalisation to High-Risk Systems 103
Postbureaucracy 104
Postregulatory State 104
Studies on Globalisation in the Field of Safety 104
Externalisation 105
Standardisation (and Bureaucratisation) 106
Financialisation 108
Digitalisation 109
Self-Regulation 110
Combining Studies 111
Back to *Normal Accident* 112
A New, *Post Normal Accident*, Operating Landscape 112
Hopkins' View and Globalisation 113
The Story of BP, the Strategic Fiasco of a Multinational 115
Browne's Legacy 116
The Networked Firm 118
A Postbureaucratic Strategy and Postregulatory State Failures 119
Globalised Trends in the BP Case 120
Financialisation 120
Digitalisation 120
Externalisation 120
Financialisation 121
Standardisation 121
Self-Regulation 121
From Component Failure to Network Failure Accidents 121
Summary of Chapter 4 123

5. (Global) Eco-Socio-Technological Systems: Expanding Scale, Scope and Timeframe 125
Introduction 125
Expanding Scope, Scale and Time Frame 126
High-Risk Systems and Sociotechnological Risks 126

Globalisation and Systemic Risks ... 127
Anthropocene, Transhumanism and Existential Risks ... 129
Widening and Complexifying the Risk Picture ... 132
Embedded Risk Categories: Complex Interactions ... 136
Loop A: Causal Circularity between High-Risk Systems and Globalisation ... 137
Loop B: Causal Circularity between Globalisation and Anthropocene/Transhumanism ... 138
Rethinking Perrow's Matrix ... 139
Scale of Governance and Magnitude of Impact ... 139
Eco-Socio-Technological Disasters ... 141
The Case of Fukushima Daïchi ... 141
General Complexity ... 142
Summary of Chapter 5 ... 145

6. Conclusion ... 147
Two Opposite Theses in One Book ... 148
Hopkins, an Unofficial Theorist for the Second Thesis of *Normal Accident* ... 148
Errors from the Top (versus Sharp-End Human Errors) ... 149
Network Failure Accidents (versus Component Failure Accidents) ... 149
From Technical to Eco-Socio-Technological Disasters ... 150

References ... 151

Index ... 165

Introduction

Post Normal Accident (Post NA) comes back on the core of Perrow's work in his seminal contribution, *Normal Accidents (NA)* (Perrow, 1984), and revisits its main insights to provide a new discourse for the study of high-risk systems. Released in 1984, Perrow's book became one of this instant classic. It defined a new era and field of investigation: the sociological study of high-risk systems and their hazardous potential. It was published in the year of the Bhopal disaster (1984), 2 years before Chernobyl and Challenger in 1986, 3 years before the Herald of Free Enterprise in 1987 and 4 years ahead of Piper Alpha in 1988.

This series of symbolic high-profile events in the 1980s in a diversity of high-risk systems (e.g. offshore platforms, space missions, nuclear power plants, aircraft) amplified the relevance of Perrow's work in the West, establishing one core controversy in a new field of research at the end of the 20th century: are man-made disasters preventable? Perrow's thesis was that it was not possible to do so for certain kind of systems because of their degree of complexity and coupling.

It was a technological deterministic argument derived from a popular philosophy in the 1970s (Winner, 1978) as described by the historian of technology Hughes (2005). Nuclear power plants were at the heart of the debate, but the book also contributed to discriminate high-risk systems (e.g., aircrafts, chemical plants, trains, mines) from other organisations (manufacturing plants, schools, etc.) and to challenge the simple idea that disaster could be explained through human error. Systems created disasters, not the process operator or pilot making a mistake. These were the 1980s.

What about *Normal Accidents* 36 Years after Its Publication?

What happened in the meantime? There have been many changes at the empirical and conceptual levels in more than 35 years. Empirically, when Perrow wrote his seminal book, the Cold War was still structuring international relations, Internet was in its infancy, finance was not yet dominating capitalism, multinationals did not compete on global markets with today's intensity and the ecological crisis had not materialised in recurring extreme events (e.g. heat waves, fires, storms, floods).

Since then, globalisation has restructured companies' environment and the role of states, finance is at the heart of capitalism and markets, Internet constitutes the core infrastructure of the digital age and the notion of anthropocene indicates that humanity now acts as the equivalent of a geological force, transforming our environment in an unprecedented scale, challenging the distinction between nature and culture.

This brief description which captures the developing situation of the past two to three decades at the world stage did not obviously structure Perrow's analysis of high-risk systems. These realities simply did not exist, although they now constitute the background of safety critical organisations. So, one can imagine very well that writing a book like *NA* today would be very different than writing it more than 35 years ago for these very reasons.

But changes are not only empirical, and conceptual ones have also been quite important in the past 36 years in safety research. Important contributions in the sociology of safety, human factors, cognitive engineering, system safety or regulatory studies have provided many new ideas to think about high-risk systems, ideas that were not available at the time of the publication of *NA* for Perrow to include them in his work (see Le Coze, 2019a). As a result, our understanding of accidents is not the same today as it was back then.

For instance, Fukushima (2011), Deep Water Horizon (2010) and Columbia (2003) which mirror Chernobyl (1986), Piper Alpha (1988) and Challenger (1986) in the 1980s have not been studied and analysed the same way. Much progress has been made in our understanding of such events in the past 36 years, particularly so with the help of other sociologists than Perrow such as Andrew Hopkins (2008, 2012) and Diane Vaughan (2005, 2011), to cite only two authors, who have added substantial insights through their research to what was available in the 1980s.

So a book like *NA* cannot be read today the same way it was read more than 35 years ago. Yet, the influence of NA has not seemed to really wane. It remains this classic that people refer to when it comes to technological accidents, high-risk systems and safety management. Because of this lasting popularity of the book, *Post NA* wishes to provide readers with an update or a revisit considering the empirical and conceptual changes between 1984 and 2020 as sketched earlier.

Why the *Post* in *Post Normal Accident*?

Why the expression *Post Normal Accident?* Adding the prefix "post" to key notions is a rather common practice in the social sciences. Quite popular examples are postmodernity, postindustrial society, postbureaucracy, postnormal science or also postcolonialism and posthumanism. With postmodernity, modernity is challenged in its foundations by questioning reason and emancipatory discourses. In the postindustrial society narrative, industrial society evolves to include services and the information economy.

Postbureaucracy challenges bureaucracies since they are replaced by network properties that restructure businesses. Postnormal science considers science to be more opened to uncertainties, to its limits and its value-laden dimension, whereas postcolonialism critically examines many of the dimensions associated with colonialism among which its European-ethnocentric discourse. And, in posthumanism, humanism is questioned by a more inclusive perspective shaped by an ecological mindset rejecting the separation between nature and culture. These examples of "post" notions have been quite successful because of their ability to address new situations while triggering debates and controversies among scholars and beyond. The prefix "post" stresses therefore that key notions that captured a moment of history in various domains seem to lose their relevance and need adaptation.

Because Perrow's seminal piece *NA* established a very strong background structuring the field of safety over 30 years, during the 1980s, it makes sense to define *Post Normal Accident (Post NA)* as a systematic discussion and critique of each of its core insights. The chapters of this book follow this "post" strategy with safety and technological disaster studies. They build up, one step after the other, a move from *NA to Post NA*.

Post Normal Accident in a Nutshell

What is the main argument of *Post NA? NA* argued that some types of disasters could not be avoided because of tight coupling and interactive complexity. Perrow successfully visualised this idea by using a 2 × 2 matrix which distinguished between what he singled out as high-risk systems, a particular kind of organisations because of the threat they pose to societies. Some were inherently more likely to surprise us according to Perrow (e.g. nuclear power plants) than other (e.g. mines).

However, there was another thesis in the book that most accidents were the results of organisations favouring other goals than safety, more so in poorly regulated high-risk systems. This was literally the opposite argument. Accidents could be prevented. It was not surprising considering that Perrow was a critical sociologist of organisations with a broad scope. Perrow introduced analytical categories to ground this second view of accidents and used the combination of technology (tasks), structure, goal and environment of high-risk systems to explain how certain properties were more favourable to safety than others.

One key insight was that safety is a question of power relationships between actors and institutions (e.g. professions, unions, civil society, insurance, justice, states, corporations) involved in the performance of these systems rather than only an intrinsic technological feature of systems. Perrow had another categorisation for accidents that happened in the systems for which the balance of power was unfavourable to safety because of failure of top management: they were component failure accidents.

Post NA recognises and integrates the importance of this second thesis, and then identifies another sociologist, Hopkins, who contributes to elaborate further this second thesis, creating over the years a normative model of safety which helps assert Perrow's second message that *accidents are normal because they repeat despite knowing that they could be prevented in principle.*

But Post NA also extends further the scope of *NA*. It argues in favour of the perspective advocated by Perrow (and Hopkins) that the empirical study of safety in high-risk systems has to incorporate explicitly an analysis of top management. *Post NA* translates this claim by arguing that strategic decision-making should be part of the conceptualisation of safety more than has been done so far, and a distinction between strategic mistake, failure and fiasco is a first proposition towards that goal.

Post NA also explores the implication of globalisation for high-risk systems by sensitising Perrow's (and Hopkins') analytical categories of technology (task), structure, goal and environment to trends such as digitalisation, externalisation, standardisation, financialisation and self-regulation. To accentuate this distinction with the past, *Post NA* complements the idea of component failure accidents of *NA* with the notion of network failure accidents (which is illustrated by the story of British Petroleum (BP)).

Post NA revisits the 2 × 2 matrix by situating high-risk systems more broadly than they were situated in the 1980s by identifying systemic and existential risks. These new risk categories translate the globalised and ecological context of our current times while tremendously expanding dimensions such as scope, scale and time frame. A new type of matrix is suggested, replacing the variables "coupling" and "interaction" with the dimensions of "scale of governance" and "magnitude of impact".

Finally, *Post NA* investigates the implication, for the core NA notion of interactive complexity, of considering such dimensions as sociotechnological, systemic and existential risks together. Complexity was a central analytical category of Perrow in the 1980s, but it needs to be revised to accommodate the scope, scale and time frame now faced by high-risk systems. The formulation of general complexity is the substitute to Perrow's interactive complexity and illustrated by Fukushima Daïchi, described as an eco-socio-technological disaster.

Book's Structure

The book is structured as follows. Chapter 1, "One Book, Two Theses", and Chapter 2, "Hopkins, the Unofficial Theorist of NA", introduce the mostly discarded second thesis of NA and its unofficial theorist, Hopkins. Chapter 3, "Error from the Top"; Chapter 4, "From Component Failure to Network Failure Accidents"; and Chapter 5, "Global Eco-Socio-Technological Systems: expanding scale, scope and timeframe", revisit some of the core insights and propositions of NA: the 2 × 2 matrix, the notion of component failure accidents, the distinction between technology (task), structure, goal and environment of organisations, interactive complexity and the importance of thinking strategy when it comes to the safety of high-risk systems, beyond human error of front-line operators.

The chapters are based on the following articles published between 2015 and 2019, Normal Accident. Was Charles Perrow right for the wrong reasons? *In Journal of Contingencies and Crisis Management.* 2015. 23 (4). 275–286 (Chapter 1); Storytelling or theory building? Hopkins' sociology of safety. In *Safety Science.* 2019. 120. 735–744 (Chapter 2); Safety as strategy: mistake, failure and fiasco in high-risk systems. In *Safety Science.* 2019. 116. 259–274 (Chapter 3); Globalization and high-risk systems. Policy and Practice in Health and Safety. 2017. 15 (1). 57–81 (Chapter 4); and An essay: societal safety and the global 1, 2, 3. In *Safety Science.* 2018. 117. Part C. 23–30 (Chapter 5). These articles have been revised, adapted and articulated for this book.

Author

Jean-Christophe Le Coze is a safety researcher (PhD, Mines ParisTech) at INERIS, the French national institute for environmental safety. His activities combine ethnographic studies and action research in various safety-critical systems, with an empirical, theoretical, historical and epistemological orientation.

1

One Book, Two Theses (NA and na)

Introduction

In 1984, Perrow released a landmark book entitled *Normal Accidents* (*NA*), in which he defended the argument of the inevitability of accident in certain type of high-risk systems. This is the first, most visible, thesis of the book which is mainly retained by its readers. The aim of this chapter is, 36 years after the book's publication, to reconsider its content and highlight its second thesis. Indeed, in hindsight, the message of the book seems today as compelling as ever, but this impression needs an update.

Accidents and catastrophes have indeed continued to occur over the past three decades, and it is one reason for *NA*'s enduring influence. However, its technological determinism has been criticised and shown to be too restricted to account for the social nature of disasters. In opposition to his technostructuralist view, some authors have argued about an extended sociotechnological models of the normality of accidents.

Indeed, Perrow himself often complained about the misuse of his own technological rational to interpret accidents. Consequently, he has regularly used instead a wider, macrosocial, critical and power view of organisations and society to interpret accidents. This second posture was clearly in the background of the 1984 book but has often been neglected, unnoticed or overlooked by commentators, proponents and critics. The purpose of this chapter is therefore threefold.

First, it offers the reader a presentation of *NA*, introducing its key concepts and not exclusively, as is often the case, the concept of normal accident (na). Second, the chapter shows that Perrow has a much wider interpretation of accidents than many of the commentators have granted the American sociologist in the heat of the debates about the inevitability of accidents. Perrow is an important author of the sociology of organisations, which provides a profound understanding of the prevalence of corporations and their mode of operating in and influencing societies.

Third, this interpretation is a new point of departure to explore further the implication of this second thesis. In the following sections, I therefore extract and amplify, retrospectively, a different message from the book *NA*.

A first reading of the book focuses on its technological rationale without taking into account Perrow's extensive sociological research. Another reading entails the explicit link of the book's content to Perrow's critical view of organisations and society developed over the past 40 years, before and after the publication of *NA*.

However, this operation can only succeed in the context of a wider perspective of the sociologist's legacy that goes beyond his interest in safety and disasters. This chapter is organised as follows. After a short introduction to the book, I consider the reception of its main thesis in the 1980s and 1990s. I then suggest considering the book from a broader perspective in order to elaborate this alternative reading and explain the existence of a second thesis in *NA*. Finally, I extract key interpretive principles that Perrow uses: cognitive, organisational and macrosystemic.

The Reception of *Normal Accident*

Singling Out and Comparing High-Risk Systems in the 1980s

Perrow's argument is one of the most popular in safety. It is one of the corner stones of industrial safety, developed at a time when the field was slowly emerging as a legitimate and independent object of scientific investigation. In a nutshell, his thesis is that in tightly complex systems, accidents are, from time to time, inevitable. One conclusion is that tightly coupled and complex systems with catastrophic potential should be abandoned or made less coupled and less interactive.

This contention was illustrated by a wealth of descriptions of accidents in different systems, including nuclear (Chapter 2), chemical (Chapter 4), aviation (Chapter 5), marine (Chapter 6), mining industries and dams (Chapter 6) and also space, weapons and DNA engineering (Chapter 7). By doing so, Perrow singled out a specific set of techno-organisations, so-called "high-risk systems", for their hazardous potential.

This helped establish and distinguish new areas and topics of research within the social sciences: technological risks, safety and disasters. And the rich text of Perrow was accompanied with a matrix which not only summarised but also further conceptualised his argument (Figure 1.1).

The top right quadrant is dedicated to systemic accident proneness (na) with nuclear plants and weapons, space missions or aircraft because of their intersection of complexity and tight coupling. Not only did this picture contribute for the first time to an identification of high-risk systems (e.g. dams, marine transport, rail transport, mining, chemical plants, nuclear plants, airways), but it also classified them according to their potential for unexpected surprises: normal accidents.

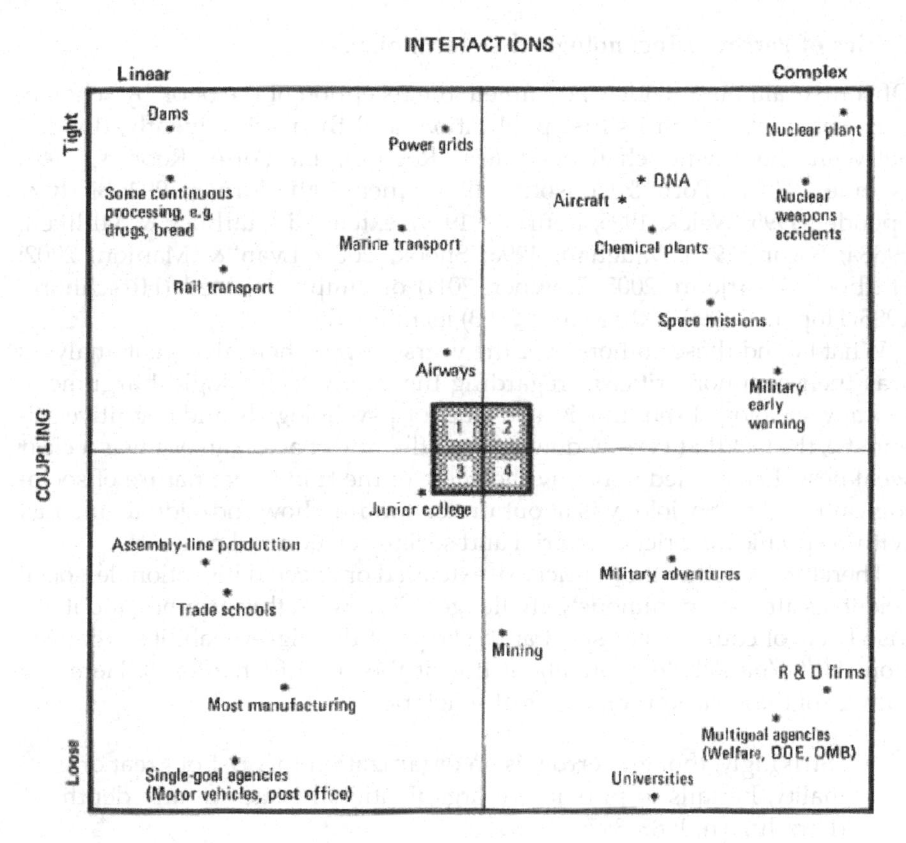

FIGURE 1.1
Perrow's matrix of high risk systems.

It brought an additional insight beyond what the text provided. It grouped together the high-risk systems that had first been analysed in the book, organising them in relation to each other, illustrating perfectly the spatial property that only drawings can afford to authors. It contributed without a doubt to its success.

Because of its quality and timeliness, as well as the centrality of its topic in contemporary debates about increasing technological developments (particularly with its strong attack of the nuclear industry), the book managed to find a cross-disciplinary readership (e.g. engineering, psychology, cognitive engineering, sociology, history).

Thirty-five years later, it remains very impressive in its breadth and ability to combine a wide range of data with a high level of conceptualisation. The fact that in subsequent years, some of the most symbolic accidents of the 1980s (e.g. Bhopal, Challenger, Chernobyl) occurred also helped to convey a special and prescient character to the book and its author.

Critics of Perrow's Technological Determinism

Of course, although widely acclaimed, the reception of the book by scholars was, sometimes from its first publication, and then subsequently, divided between those who challenged (e.g. Rochlin, La Porte, Roberts, 1987; Wynne, 1988; La Porte & Consolini, 1991; Pinch, 1991; Turner, 1992; Bierly & Spender, 1995; Weick, 1995; Bourrier, 1999), extended (Starbuck & Milliken, 1988a; Sagan, 1993; Vaughan, 1996; Snook, 2000; Evan & Manion, 2002; Starbuck & Farjoun, 2005; Downer, 2011) or simply rejected (Hirschhorn, 1985; Hopkins, 1999, 2001; Mayer, 2003) its rationale.

What bound these authors over the years despite their divergent analyses was their common criticism regarding the overly technological argument. For any sociologist (but this is also true for psychologists and cognitive scientists), the fact that Perrow downplayed the role of actors in *NA* was a clear weakness that seemed to betray the value of the traditional nature of sociological insights. Sociology is about understanding how individuals interact within specific historical, material and socially structured contexts.

Therefore, whether they criticised, extended or rejected his rationale, social scientists almost unanimously challenged Perrow on this ground (and it has also been, of course, one aspect at the heart of the high-reliability organisation (HRO)/na debate, more about this at the end of Chapter 2). Here is a sample of these various reactions of scholars:

- "Strikingly, though Perrow is an organization theorist of great originality, he fails to pursue the organizational issues in any depth" (Hirschhorn, 1985, 847).
- "Despite being trained as a sociologist, Perrow essentially ignored the people manning and completing the high risk system, and the human tendencies which lead to socialization" (Bierly, Spender, 1995, 645).
- "Perrow's marginalization of human, organizational, and socio cultural factors surrounding the causes of technological disasters leads to an impoverished understanding of the complex dynamics of technological disasters" (Evan & Manion, 2003, 89).
- "The Challenger disaster can justifiably be classed as a normal accident: an organisational-technical system failure that was the inevitable product of the two complex systems. But this case extends Perrow's notion of system to include aspects of both environment and organisation that affect the risk assessment process and decision making" (Vaughan, 1996, 239, 415).

It is clear that the core of his thesis lies deep in technological determinism, as he himself states clearly throughout *NA*: "Besides being a book about organizations (but painlessly, without the jargon and the sacred texts), this is a book about technology. (…) As the saying goes, man's reach has always exceeded his grasp. (…) Systems are too complex, and our reach has surpassed our ability to grasp. There is a bit of this philosophy in this book" (Perrow, 1999, 11, 339).

I have as a consequence suggested elsewhere (Le Coze, 2018) to characterise this view of the normality of accidents as the "Ellulian" thread, in reference to Jacques Ellul, the social historian and thinker who was one of the first to describe technology as evolving out of human control (Ellul, 1954), but also in contrast to other versions of the normality of accidents (I come back to this a bit later). Disasters will happen in certain systems even when everyone is working to avoid them.

Component Accidents versus Normal (or System) Accident

But, very quickly after the publication of his *NA* thesis, for example following the Challenger explosion (Perrow, 1986b), Perrow started to warn against what he could see as a misunderstanding of the central idea laid down in the now classic book. Accidents do not systematically happen because of intrinsic and structural features of high-risk systems.

Managerial and/or regulatory issues explain many disasters, perhaps even the majority of them. What he had qualified as component failure accidents (namely, accidents that do not result from many independent failures) was a category especially created to distinguish the two types of events, identified as issues in the DEPOSE system of companies (design, equipment, procedures, operators, supplies and materials, and environment).

Thus, there are many component failure accidents that could be prevented. For the sociologist, the Challenger is not a normal (or systemic) accident, and Bhopal and Chernobyl are not quite either, as this quote attests: "Following the publication of the book in 1984, we had a series of notable accidents: Bhopal, Chernobyl, and in the United States, the Challenger and the Exxon Valdez oil spill. I was told I had bragging rights, but this was not true. All of them were just waiting to happen: the result of managerial negligence and, in addition, in the case of the Challenger, of extreme production pressures. Normal accidents are fairly uncommon and, I add, cannot be completely prevented. These accidents were all preventable. The vast majority of accidents are preventable" (Perrow, 1999, 23).

Hopkins' Refutation of *Normal Accident*

Although Perrow stressed the centrality of this distinction and insisted that many accidents were component accidents according to his terminology which had to be analysed as the consequences of managerial and regulatory failures, the Australian sociologist Andrew Hopkins moved on to produce a persuasive and determined attempt to disqualify Perrow's technological determinism at the root of *NA*. For this purpose, several convincing rebuttals were assembled in a series of two articles (Hopkins, 1999, 2001). These rebuttals included the following facts:

- First, *NA* does not apply to many (if any) disasters.
- Second, there are certainly in the book many valuable insights on accidents, but they don't actually fit the technological argument.
- Third, organisational sociological concepts are not an integral part of this argument.
- Fourth, it is difficult to classify systems unambiguously along the two features of coupling and interaction.
- Fifth, HRO studies showed the possibility of combining modes of organising compatible with the paradoxes of managing safety in tightly coupled and complex systems.
- Sixth, Three Mile Island (TMI) in 1979, used as the paradigmatic example of *NA*, was not a valid illustration. TMI can, instead, be more fruitfully analysed from the perspective of a (sloppy) management failure to communicate information rather than an example of an inevitable accident.
- Finally, Hopkins also considered that the initial and laudable motivation of Perrow to avoid a narrow interpretation of accidents through "human error" can be dealt equally well with the help of other accident models (e.g. Turner, 1978; Reason, 1990).

Despite the fact that this appears at first glance as a quite compelling refutation of *NA*, this refutation can also, in turn, be reappraised. It is particularly interesting to do so because the two papers (Hopkins, 1999, 2001) are quite systematic in that they included what were then scattered criticisms to which many other authors might certainly subscribe (see samples given earlier).

One dispute with this refutation is that it amounts to the condemnation of the entire book by reducing its wealth of data to Perrow's structural (coupling/interaction) framework, as if the rest of the insights collected in the book had no coherence. This was not said specifically in Hopkins' articles, but was at least implicit, without the option to explore the opposite possibility that there was another (partly hidden but consistent) rationale beyond the technological one.

This framing used for refuting *NA* isolates the book from Perrow's wider contribution to sociology. Many other critics have also proceeded similarly. It is a defendable strategy, and I think entirely justified. But this, as the next sections will show, is problematic because it is difficult today, retrospectively and considering Perrow's subsequent reactions to disasters over the past 30 years, to separate the book from its specific critical (and widely acknowledged) position within the field of (organisational) sociology.

In the light of this retrospective stance and broader perspective, many of the interpretations of the book that indeed could be seen as external to the main argument appear nevertheless to make much more sense. This quote illustrates it: "NAT is propounded in a book, NA, which is full of interesting insights essentially unconnected to the theory" (Hopkins, 1999, 101). Yet, when one analyses and situates the book historically and conceptually in relation to Perrow's broader intellectual production and wider reception as well as his own additional input to *NA*, it is indeed possible to obtain an alternative and complementary approach to the book.

Hopkins' Omission: The Extended Version of Normal Accident

But another interesting omission in Hopkins' critics of *NA* is the absence of a mention of the extended versions of na first by Vaughan (1996) with *the normalisation of deviance* and then *the dark side of organisation* (Vaughan, 1999), which Hopkins clearly knows of (Hopkins, 1999), and a second extended version by Snook (2000) with the *practical drift*, which he might not know back then (one cannot find it mentioned in any of Hopkins' writings of the late 1990s and early 2000s). Snook writes, "this accident was "normal", not only in the sense that Perrow suggests – "that it was an inherent property of the system" (1984:8). But rather it was normal because it occurred as the result of normal people behaving in normal ways in normal organizations" (Snook, 2000, 202).

If fact, if one wants to be more accurate, both conceptually and historically, the na idea, namely that accidents happen in high-risk systems despite efforts to prevent them, was also discussed in parallel, sometimes even earlier than Perrow, but using different conceptual lenses. Turner and the *incubation period* (Turner, 1978) insisted on our cognitive, bureaucratic and cultural limitations to anticipate (a "Kuhnian" thread, Le Coze, 2018). Rasmussen with the *defence in-depth fallacy* (Rasmussen, 1990) emphasised instead the adaptive nature of complex behaviours in distributed high-risk systems and their potentialities for unexpected patterns of interactions (an "Ashbyan" thread, Le Coze, 2018).

The core of the *dark side of organisation* model which frames Vaughan's research (Vaughan, 1999, 2011) is based on the attempt to conceptually connect literatures on routine nonconformity, mistakes, misconduct and disasters through environment of organisations, organisations' characteristics and individuals' practices (cognition and choice). Vaughan contrasts Perrow's structural and technologically based framework with her own by emphasising individual agency, interaction and social structure. In Vaughan's model, "(1) these are systems of human interaction, so how actors interpret the system and strategize is significant; (2) structure is strongly influential, but not fully determinant: agency and contingency also are important; and (3) time, history, and the trajectory of actions and interactions matter" (Vaughan, 2005, 42).

This description is compatible with Snook's analysis which articulates analyses at the individual, collective and organisational levels into a broad pattern of drift. These alternative and extended versions of the concept of na are more explicitly social than Perrow's initial version. With Perrow, it is the combination of tight coupling and complexity which creates the potential for surprises; with Vaughan and Snook, this idea migrates to the cognitive, organisational, cultural, social and political complexities of networks and systems. The pessimistic tone of the original na idea is kept; events in high-risk systems are very likely to surprise us from time to time, but it is with a more explicit connection to a more sociologically oriented type of analysis.

Such an extended version of na could be summarised as follows:

1. High-risk systems face internal and external complexities, constraints and pressures (e.g. technological, social, political and economic), and as a result, operators, engineers, managers, executives and regulators interact and deal with a wide range of uncertainties, leading to innumerate, local and simultaneous decisions across hierarchies, functions, structures and social differentiation of expertise and tasks in multiple organisations.
2. Because of the complexity of these interactions within these socio-technological systems (and networks), no one can predict with certainty their behaviour, no one can pretend to produce the big picture that would help entirely describe these interactions to obtain full control of them, and the imperfections of organisations have become accepted (and expected) by many.
3. As a consequence, the boundary between operating safely and operating unsafely (or between the dark and the bright sides in Vaughan's vocabulary) is probably more ambiguous that many (from private companies to regulators) are often willing to publicly admit in a world that promotes technological, managerial and financial innovations translated in (sometimes) ambitious companies' strategies.

For these writers, accidents can be the product of the complex interactions of technological, organisational, cultural, social and political dimensions. Despite everyone's good intentions, accidents can nevertheless happen. "No fundamental decision was made at NASA to do evil, rather, a series of rather seemingless decisions were made that incrementally moved the space agency toward a catastrophic outcome" (Vaughan, 1996, 410). These are important and contentious claims for many, including Perrow, who has a rather ambiguous stance towards these kinds of additions to *NA* (but also Hopkins, more about this sociologist in Chapter 2). Perrow, for instance, disagrees with Vaughan and warns, "we miss a great deal when we substitute culture for power" (Perrow, 1999, 380), a sentence that one understands much better with a more in-depth perspective of his work.

Charles Perrow's Sociology

Seven Related Books

In this section, I return more broadly to Perrow's contribution to sociology, and more specifically to his (macro)sociology of organisations. I proceed first chronologically. Next, I describe a pattern of interwoven interests revealing the consistency of a lifetime of research in social sciences and the topic of organisations. In the 1970s, Perrow published three books: *Organizational Analysis. A Sociological View* in 1970, *The Radical Attack on Business. A Critical Review* in 1970, and *Complex Organizations. A Critical Essay* in 1972 (with two new and extended editions in 1979 and 1986). Then, in the 1980s, he published *Normal Accident, Living with High Risk Technologies*, in 1984. *The AIDS Disaster, The Failure of Organizations in New York and the Nation* was published in 1990, in collaboration with Mauro Guillén.

The next books were released in the first decade of the 21st century, starting with *Organising America. Wealth, Power and the Origins of Corporate Capitalism* in 2002, followed by *The Next Catastrophe. Reducing Our Vulnerabilities to Natural, Industrial and Terrorist Disasters* in 2007 (with a new edition in 2011). In between the publication of these books, he published a very significant number of papers and articles that anticipate, summarise or develop the major themes explored in the various books. Some of these articles prove to be very valuable for the argument developed here.

My goal is to consider Perrow's basic analytical principles in sociology that shed light on his legacy for the field of safety and help us surface the second thesis of *NA* that is the subject of this chapter. In this respect, one first thing to notice is that risks, accidents, disasters and safety are the subject of less than half of all of Perrow's books, which indicates that there is more to learn about his ideas if one looks beyond these concerns.

This is a very interesting situation, because our view of safety and accidents is influenced by personal convictions regarding many topics (technology, organisation, society, progress, democracy, etc.) that are often beyond the presentation and interpretation of data and remain implicit. In this case, it is the opposite; the wealth of Perrow's writings allows the readers to infer quite well many of his personal convictions.

A Set of Interwoven Studies

There are, I believe, two other major subjects in Perrow's research apart from disasters and safety: organisation theory (Perrow, 1973) and the theme of the "society of organisations" (Perrow, 1991). The former is elaborated in two books (and related articles): one in which Perrow describes his sociological approach to organisations (*Organizational Analysis. A Sociological View*, 1970) and the other in which he offers a critical appraisal of the history and theories of the field combined with a refined version of his own theoretical input (*Complex Organizations. A Critical Essay*, 1972).

The latter, the theme of the "society of organisations" which concerns a macroview of society, is analysed from a sociopolitical angle in a book entitled *The Radical Attack on Business. A Critical Review* (1970) and from a historical perspective of the emergence of capitalist organisations in the 19th century (*Organising America. Wealth, Power and the Origins of Corporate Capitalism*, 2002).

These two themes are strongly connected because it is on the basis of a specific approach to organisations established within the field of organisational theory during the 1960s and 1970s that he explores empirically the emergence of society of organisations in the 1980s and 1990s. And it is, obviously, partly on the basis of all these developments that he offers his insights on safety and disasters in the 1980s and in the years to follow.

It is worth noting that his interest in accidents began, as can be inferred from his writings, when he answered positively to the request to participate and offer a social scientist view to the presidential report in the aftermath of TMI. Hence, one finds this comment in the preface of *NA*: "At the time of the accident I was researching the emergence of organizations in the nineteenth century. I still hope to return to that comparatively sane world" (Perrow, 1999, viii).

But disasters turned out to be very good examples and case studies of organisational "externalities", to employ the terminology of Perrow, namely the consequences that organisations generate for society without bearing their costs (e.g. pollution, transport needs, urban concentration). The topic of accidents and disasters has since become a good opportunity to apply his organisational knowledge, as expressed in the third edition of his critical essay on organisational theory.

"Organizational analysis will realise this promise, I believe, if it can be shown to illuminate such important public policy questions" (Perrow, 1986a, 154). From this point of view, it is possible to understand Perrow's attention to technological risks as an occasion to illustrate but also to apply or test empirically his personal concept of organisation and society.

In this respect, what he wrote in the 1970s on organisation theory applies very well to the *NA* thesis and the high number of case studies which are its building blocks. "Since conceptualization – ways of looking and seeing – should not proceed in a vacuum, I have made a considerable effort to include descriptive and anecdotal material. One might even gather that I spend most of my time reading company stories in Fortune. A good way to discipline theory is to relate it to descriptions made of other purposes, even though this is often considered mere 'illustration'. Furthermore, I know no better way to present the flavour of organizational life than by dealing with actual organizations" (Perrow, 1970b, viii). From this bibliographic introduction, one can begin to contemplate a very interesting set of interwoven research topics in Perrow's intellectual production: (1) organisation theory, (2) society of organisation and (3) safety, risk and disasters.

From this perspective, it appears clearer that there could be, at least, three strategies to investigate *NA*:

1. isolating the book from previous and subsequent works
2. interpreting the book in the light of earlier ones
3. looking back at the book from the present and including all previous and subsequent writings to *NA*'s publication

As argued earlier, the first strategy is limited, and thus, a choice must be made between the two other approaches. It is the third that is selected here because it is the most complete. Perrow has continuously refined his view on accidents in the years following its publication in 1984, introducing the notions of executive failure or malfeasance, whether in books (Perrow, 2011b) or in articles, the last ones on the Fukushima Daïchi accident being particularly revealing (Perrow, 2011a, 2015).

The second strategy will, however, occasionally be used to strengthen the message of this article. Thus, three major research topics, namely, organisational theory (1), accidents and safety (2) and the society of organisations (3), are taken to be interconnected topics. Perrow himself has recently commented on the common thread between many of his books (Perrow, 2012), especially *NA* (1984), *Organizing America* (2002) and *The Next Catastrophe* (2007).

He defends a more decentralised or networked type of society, in which organisations would be smaller and able, first, to reconfigure themselves when facing external stresses (whether economical, natural or terrorist), second, to limit the accumulation of economic and political power, and third, to limit the concentration of technological risk.

My interest lies in a complementary view. To do so, I need to describe both Perrow's critical approach to the sociology of organisations, based on a tool and power view and his macro-organisational sociology of the society of organisations. From there, I will be in a position to explain how one can extract a different message from *NA*, with the support of his other articles and books written on this topic.

Perrow's Sociology of Organisations

The Context of the 1960s

Perrow is a leading theorist in the field of organisations. He was a student of Philip Selznick, who was himself one of the founding fathers of the sociology of organisations, along with authors such as Alvin Gouldner, Peter Blau and Michel Crozier. Perrow's historical and critical account of the field of organisational theory is a classic (Perrow, 1986a), no less than Richard Scott's several editions of his introductory manual in the field of organisations (Scott, 2003). For Perrow, as for these pioneers, Max Weber's insight on bureaucracy remains the classic figure for a sociological investigation into this area.

In contrast with current trends in organisational theory over the past 20–30 years, which tend to leave aside many of the concerns of these early sociologists who were able to situate more widely the implications of the rise of bureaucracies, Perrow has maintained a clear empirical focus on trying to situate the importance of organisations to our understanding of society through a critical eye, and a power perspective, targeting the role of elites.

Thus, this situation and the influence of Perrow have been fairly well stated in the field of organisation theory: "Selznick (1949), Gouldner (1954) and Perrow (1986) all point to these issues. Particularly, they emphasize that the actual behavior of organizational members is determined more by the conflict between opposing factional interest within an organization, or between those factional groups and such groups outside an organization, than by any overarching goals or unified, legitimate structures. Indeed such goals and structures are the work of powerful organizational members who have the ability to design structures and systems and manipulate incentive systems. Organizations are arenas of conflict, with senior managers able to achieve dominance" (Hinning & Greenwood, 2002).

And, it is precisely this ability to keep an eye at the same time on organisations, elites' interests and society as a whole that, in my view, confers to Perrow's legacy so much relevance for the field of safety. I wish now to attempt to explain what I believe to be Perrow's principal ideas about organisations, although this of course will simplify many of the nuances necessary to give full credit to the subtleties of his writings. The intention here is not to be thorough but explicit enough for the purpose of this chapter.

It is relevant to notice that in the 1970s and 1980s, his writings on organisations seemed to be based on four motivations. The first was to distinguish the sociological view from the psychological, economic, political or administrative ones. The second was to develop this first point in the clearest manner possible by explaining how it differed from mainstream approaches in the 1960s and 1970s, including the human relations school.

This quote summarises how he proceeded, by contrasting his key sociological variables from those belonging to different disciplinary or research traditions. "Structure, power, goals, environment, these are the concepts that have been stressed in this book. Leadership, interpersonal relations, morale and productivity – these concepts have not been stressed. The difference is one of perspective" (Perrow, 1970b, 175).

The third is within the narrower field of the sociology of organisations; elaborating a critical stance in order to reorient a dominant thesis, i.e. the institutional and neoinstitutional interpretation that emphasises symbols and myths (or culture). The following quote is exemplary of this project, which comments about this prevailing version: "It is also a product of the guiding perspective of sociology since the 1930's – value free functionalism, the pluralist doctrine, and the emphasis of norms, values and culture at the expense of power and the material aspect of existence (...)" (Perrow, 1986a, 177).

A final motivation in his writing is the need for theoretical developments that would allow social scientists and observers in general to easily understand organisations with their diversity and to make sense of the abundance of information about them that one can find in books, articles, newspapers and personal experiences. "What we must discover are patterns of variation, which hold despite the uniqueness of markets, structure, personnel, history, and environment and which provide fairly distinct types that can be used for analysis and prediction" (Perrow, 1970b, 177), and "The social sciences, at least in the area of complex organizations, have been desperate for ideas, not data" (Perrow, 1986a, 83).

Four Key Dimensions: Technology, Structure, Goals and Environment

It is now apparent that Perrow built his own distinct view to organisations by contrasting it with other approaches. Let's return to this question to better situate his contribution. First, without rejecting all efforts directed at improving the life of organisational members, he was particularly reluctant to accept that analysis and programs targeting isolated individuals, through psychological or social psychological lenses, would be of great help to explain or to substantially modify organisational behaviour (including morale or productivity, two of the central research topics in the 1960s). As a sociologist, he identifies other dimensions that afford more powerful understanding of the situations encountered in organisations, starting with technology and structure.

Technology (Task) and Structure

As a proponent and actor in the 1960s and 1970s of the so-called "contingency theory" (Lawrence, Lorsch, 1967; Perrow, 1967), Perrow saw the diversity of organisations as a result of their adaptations to the characteristics of the technological processes designed to produce their outcomes. Therefore, structure, another key dimension along with technology (and tasks) and derived from the Weberian description of bureaucratic features (specialisation, formalisation, centralisation of control, standardisation and hierarchy), is central to any appreciation of organisations, and they must be adapted to their technological core processes and tasks.

But beyond this and most of all, the start of any sociological endeavour in the field is to recognise that bureaucracies have proved to be perfect unobtrusive means of exercising power over individuals in the context of the emergence of organisations in the 19th century. These bureaucratic structures allow owners to restrict variability in the behaviour of their members, to organise decision-making processes and to shelter, to a certain extent, organisations from external influences by standardising expected behaviours of their members (e.g. irrespective of their origins).

Of course, they are also, in many instances, clear limits to this picture of organisations when one observes their dysfunctional outcomes, informal side and unanticipated consequences, partly as a consequence of the cognitive constraints, as understood since the ground-breaking contribution of Herbert Simon regarding our bounded rationality (Simon & March, 1958).

But their dysfunctions and informal aspects exist because organisations always serve the interest or goals of, in the vocabulary of Perrow, their masters (but also any other group interests in a position to gain influence). "Organizations are tools designed to achieve various goals (…) for both the social scientist and his management trainee, the most complete understanding of an organization will come through an analysis of its goal and basic strategies" (Perrow, 1970b, 180).

Goals

To stress the goals and interests of members of organisations, and especially leaders (or masters) along with their informal and dysfunctional nature, leads to an important piece of methodological advice from Perrow: "before assuming that an outcome was unintended, it is best to see if someone in top management might not have had reason to have intended the outcome. Bureaucracy is not the breeder of incompetence, as we often would like to believe. Instead, bureaucratic organization allows leaders to achieve goals, some of which are unannounced and costly for the rest of us and are only attributed to incompetence" (Perrow, 1986a, 13). But this rule for the analysis of organisations has to be used with caution, this topic being particularly complex, and "organizations are tools in the hands of their leaders, but they are imperfect, not completely controlled, tools, and it is a struggle to maintain control over them" (Perrow, 1986a, 134).

Environment

Moreover, organisations still remain, despite various attempts, open to the influences of their environment, but, against the institutional and neoinstitutional theories, they also contribute, and in particular the governing elite, to shape them. Indeed, "symbols in particular and culture in general are not politically neutral; they can be created and propagated by political and economic elites. In a society of organizations, a theory of organizations is needed to appreciate this fully" (Perrow, 1986a, 269). This environment can be understood through several layers of investigation, including networks, sectors, industries, states and wider cultural society levels.

The theoretical result of all this is a rather complex, critical and quite nuanced but very distinct perspective on organisations that combines technology (tasks), structure, goals and environmental questions interwoven with issues of power as well as culture. These four dimensions are now essential ingredients of any introductory book to the sociology of organisation (e.g. Scott, 2003) and organisational theory (Hatch, Cunliffe, 2013).

What one can retain from these books is a multidimensional construct that requires observers to be aware of the complexity of organisations, without falling into the trap, for Perrow, of a functionalist/pluralistic view that neglects to place power asymmetries in the context of wider societal stratifications and relationships between legal, political, and economical spheres. This aspect is the topic of the next section.

The Society of Organisations

A Sociological Tradition: Karl Marx and Max Weber

"Max Weber was right in giving so much attention to organizations as the central phenomena of modern society. Indeed, there was probably a brief time in the 60's when organizational scholars thought that if we studied all societal institutions from an organizational perspective (e.g. education, religion, health, politics), then we would quite quickly have most of the answers to understanding the functioning of society!" (Hinning & Greenwood, 2002). This is, without a doubt, the central idea of Perrow's social science research throughout his career.

And this is what is particularly interesting here. Although his main area of empirical investigation is organisations, it is never at the expense of a truly macrosociopolitical prospect. I think that Karl Marx and Max Weber are most likely the two classic sociologists who shaped his sociological stance. This concern is visible very early on. Thus, in *The Radical Attack on Business* (1970), Perrow offers a sympathetic analysis of the activities of the New Left, a political alternative to the (old) Left, during the 1960s in the United States.

What is interesting in this book and the analysis therein is Perrow's firm intention to retain the sociological ambition of the founding fathers of the discipline, indicating in some places the core tension between the critical and politically engaged vision of Karl Marx and the opposite value-neutral ambition of Max Weber, both of whom addressed society as a whole, as a system.

> One of the many things that distinguished Karl Marx was his attempt to 'put it all together,' to define every significant area of life in terms of the relationship of classes of men to the factors of production. (...) The impressive sociology of Max Weber, who wrote at the turn of this century, was sometimes explicitly a dialogue with Marx. Weber insisted that status was independent of class, and so was power. He pulled it apart. Society is a system, of course, but there is no overriding case that economic processes defined it. Contemporary social science has made the disintegration of a one factor view of the system a virtue and called it pluralism.
>
> *Perrow (1970a, 267)*

Against Functionalism and Pluralism

When *The Radical Attack on Business* was written in the 1960s and 1970s, there were indeed many good reasons for sociology to challenge the contemporary system (functionalist/pluralist) interpretive framework, advocated by the figure of Talcott Parsons (and Robert Dahl, for the pluralistic view in political science). Talcott Parsons promoted a type of sociology where stability, neutrality and global coherence of society were seen as an organic whole that downplayed conflicts, critical views and transformative processes of society. In the 1960s, these ideas had to confront major events in the United States.

Social and anticapitalist movements, political opposition to US foreign policy, economic shifts toward services and mass consumption through powerful and large productive and technologically based business corporations all challenged the functional orthodoxy of sociology and gave many ample reasons to explore alternatives to functionalism and pluralism. It is precisely the "advanced stage of corporate capitalism, the transition from problem of production to problems of consumption, from a production economy to a service economy, from a concern with widespread poverty to a concern with the quality of life" (Perrow, 1970a, 242), which engaged Perrow when scrutinising the New Left's ideas.

A key difference with functionalism (or pluralism) is how the system can have negative consequences, sustain inequalities, maintain identified dysfunctions, waste resources, etc. all while serving the interests of a few at the expense of many. Such an approach must remain macro, systemic, without subscribing to a neutral view of society. For instance, he offers, as an illustration, a critical interpretation of the problems encountered in healthcare in the United States in the 1960s.

They resulted from an interlocking set of business interests of many organisations including the drug industry, hospital equipment and hospital supply, the Blue Cross Blue Shield bureaucracies and insurance companies, the American Medical Association, the professional associations of hospital administrators, the prestigious university medical complexes, the large foundations, and the appointed and elected officials. He warns that "the prevailing viewpoint is not one that attempts to link such diverse phenomena, because the prevailing paradigm is one of pluralism. The importance of the view that business is a system, and a system that dominates and incorporates all important aspects of our lives, is suggested by this brief example" (Perrow, 1970a, 244).

Organizations as a Key "Variable" in Society

What is interesting in this description of American healthcare in the 1960s is the intention, first, to investigate at a systemic level of interpretation and, second, to look at it through the lenses of organisations and corporations. It is through interacting bureaucracies that one can understand a global pattern such as one encounters in society. This pattern is a product of the interactions between many different organisations that often serve, in Perrow's tool view, the interests of the masters. Therefore, bureaucracies, especially the most dominant, retain distinctive and explanatory features of our societies (and many, through a critical stance, of its problems).

This is something that he stated clearly: "my own inquiry is frankly imperialistic. I argue that the appearance of large organizations in the United States makes organizations the key phenomenon of our time, and thus politics, social class, economics, technology, religion, the family, and even social psychology take on the character of dependent variables" (Perrow, 1991, 725). Why is this so? For Perrow, organisations have considerably shaped our contemporary lifestyles to the point that they have become a main source of the understanding of contemporary social phenomena.

This is why this claim is imperialistic. It places organisation at the heart of society. Hinning and Greenwood recall that "there was probably a brief time in the 60's when organizational scholars thought if we studied all societal institutions from an organizational perspective (e.g. education, religion, health, politics), then we would quickly have most of the answers to understanding the functioning of society!" (Hinning & Greenwood, 2002).

Perrow would obviously be more nuanced than these claims from the 1960s, but he retains this basic idea and argues for a much more explicit use of the organisational "variable" in macrosociological explanations. "Big organizations are not everything, a lot of power exists outside them, and I do not try to explain everything through them. My account is not intended as a monocausal theory, though the relentless search for the organizational cause will give it that tone" (Perrow, 2002, 8). The reason is historical and rather simple. There was a time when organisations were not as prevalent as they are now. They did not play the role that they progressively started to play in the 19th and then during the 20th centuries.

Three major aspects have to be taken into account concerning the work of Perrow in this respect, which I shall now enumerate. Wage dependency, factory bureaucracy and externalities are the three transformations brought to society by the rise of organisations. It also went along with a centralisation of power in the hands of their masters, constituting a new kind of elite, a situation that differed according to the presence of a weak or strong state.

"(A) weak state will allow private organizations to grow almost without limit and with few requirements to serve the public interests; and (b) private organizations will shape the weak state to its liking (this requires state action, in the form of changing property laws)" (Perrow, 2002, 217). A political, legal, cultural and political moment in history favoured the rise of the society of organisations, with its many consequences, including disasters.

A Summary of Perrow's Sociology

Perrow provides a complex, critical but nuanced, power and material interpretation of organisations that relies on their main dimensions such as technology (tasks), structure, goals and environment. Technology constrains tasks and structure, structure constrains employee behaviour in unobtrusive ways, goals define structure, and environment shapes organisations which shape them in return. Following Max Weber, Perrow sees corporations and bureaucracies as central defining features of the 19th and 20th centuries.

Following the critical stance of Karl Marx, Perrow sees them as tools in the hand of society's elites (Perrow also keeps in sight the Marxist's material side of existence). As a sociologist, he therefore preserves a macroperspective that situates these organisations in relation to society as a whole, as a system. Organisations shape society (through wage dependency, externalities, and bureaucracy) as much as it shapes them in turn.

Private organisations intentionally shape their natural, political, legal and social environment to thrive. They are tools in the hand of their masters and can serve their interests well, although organisations remain recalcitrant tools for them. Organisations centralise resources and can exert great political and economic power. Thus, any sociological analysis of organisations and society has to be a power (but also material) analysis, in which the neoinstitutional emphasis on symbols, myths and culture (an influential orientation of the sociology of organisations, Scott, 1995) must not mask the reality of elites' interests.

Normal Accident, the Second Thesis

The two previous sections highlighted the key aspects of Perrow's overall stance in order to prepare the ground for the second reading of *NA* that is the main subject of this chapter.

Reintroducing the Book

After the publication of *NA* in 1984, over the years, many have criticized its overly technological rationale as was introduced earlier in this article. The source of normal accidents[1] (na) was the complexity and the degree of coupling of systems. This was indeed the main thesis. But there was also a second thesis consistent with Perrow's organisational and critical posture as described in the previous sections. This critical posture and second thesis of the book has gained strength in the following years, during which Perrow commented on disasters, and more recently, for example, the case of Fukushima Daïchi (or the last economic crisis).

In this section, I extract a certain number of conceptual principles that do not reveal a technological determinism but a different rationale instead quite consistent with Perrow's more global stance as a sociologist. I have classified them into three different categories which correspond to his analysis

[1] I now distinguish *Normal Accidents*, the book, *NA*, from normal accident, one of the concepts of the book, na.

of organisations (i.e. technology and task, structure, goal and environment): first, cognitive (task), second, organisational/managerial (structure and goal), and third, social, political, economic and cultural (environment), which might also be qualified as a "macrosystem view" category.

All of these categories not only anticipate trends that have been explored in the field of safety since but also provide a series of intertwined conceptual and analytical principles of great significance that go beyond the technological argument that is often singled out from the book and that could be argued to be a reductionist vision of Perrow's contribution.

First, it is important to acknowledge and recall that the wealth of descriptions of accidents in the book is impressive. In a short period of time, Perrow managed to put together a wealth of material and organized it in such a way as to be accessible to a wide readership. The diversity of industrial domains covered (nuclear, marine, etc.) as well as other areas of investigation makes it a unique case of a comparative study that has yet to be matched in this domain.

Given the extent of the data available today and also the degree of specialisation of the different disciplines involved in the field of safety and risk, it would probably be difficult to apply the same strategy in a restricted period of time. To my knowledge, it is the only attempt to compare so many high-risk systems. To obtain this outcome, Perrow used a diverse, wide and rich documentation.

This included not only accident reports from various institutions responsible for investigations (the National Transportation Safety Board, the Coast Guard) but also studies from other entities (e.g. National Aeronautics and Space Administration [NASA], Federal Aviation Administration), newspapers, articles in different scientific journals, as well as books and reports from a variety of scientists in many different disciplinary domains (cognitive engineering, sociology, political science, etc.). As he acknowledges in his preface, his students assisted him in this task, including Lee Clarke, who was to become an important sociologist in this field. But these were very early days of safety and risk research.

Sources of Tension When Writing *Normal Accident*

Second, from this bulk of information, Perrow developed a narrative structure that was intended to demonstrate the plausibility of the thesis of *NA* and its strong technological side. But it has proven to be quite a difficult exercise to force the data into the framework, and also a quite difficult strategy considering Perrow's critical view of organisations and their masters (see the previous section).

These, therefore, created a certain number of tensions that are visible and obvious in many parts of the text. In fact, one wonders, while reading the book, if the initial idea, in combination with the rather short time available (3–4 years), contributed to make the writing exercise quite a difficult one.

From what Perrow wrote in the preface, it seems indeed that *NA*'s thesis, the possibility of na, produced in the early 1980s remained the guiding principle throughout the writing process.

"I produced a forty-page paper on time, and in it where the essential ideas for this book. If only books came as fast as ideas (...) producing the book took another three and a half years with delays" (Perrow, 1999, vii–viii). Let's be reminded that this idea was developed from his interpretation of the TMI accident in 1979 of which he wrote, before extending it to other systems, that "the accident was 'normal' because in complex systems there are bound to be multiple faults that cannot be avoided" (Perrow, 1982, 173).

Thus, one source of tension is apparent when the framework fails to sensitise the rich texture of Perrow's narrative, for instance, his macrosystem view of high-risk industries. An example of this can be found in the following quote concerning marine transport: "Though notions of complexity and coupling will help us, the principle of an error inducing system is perhaps more important (...) the very notion of system accident loses some of its distinctiveness here" (Perrow, 1999, 176). In this quote, he admits that the framework is not useful for making sense of the reasons why accidents occur in marine transport; for this, a much wider macrosystem view is required. I will come back to this point later.

Another source of tension is when Perrow struggles with his own appreciation of the nature of accidents, whether they are normal, according to his technological idea or rather the product of organisational failures, what he identifies as issues in the DEPOSE system (design, equipment, procedures, operators, supplies and materials, and environment). Sometimes, it is fairly clear that "this was not a system accident, components failed because of improper management, which included taking calculated risk" (Perrow, 242).

It indicates that the framework does not help to understand the case and that we are dealing with a different kind of accident. Sometimes, it is much more ambiguous, and the tension is much stronger: "fairly gross negligence and incompetence seem to account for this accident, but I would resist that conclusion" (Perrow, 1999, 111). In this case, the na rationale prevails over the gross negligence assumption, but it is not always very straightforward to know why Perrow would favour one over the other. I will also return to this issue later, when introducing other sociological work in safety.

Beyond the Technological Rationale

So, if one looks closely, there are many indications that the coupling/interactivity framework provides limited support for sensitising the diversity of cases that Perrow introduces (from which derives what has been called "tensions" in many places), something that he was quite aware of. I imagine that it was something that he accepted as part of his conceptualisation and as part of a rather new and quite complex field of scientific exploration that he contributed to shape. Two indications are available in his writings to corroborate this claim. Both of them are found in 1986, only 2 years after the release of *NA*.

The Case of the Challenger

The first one is his position on the Challenger accident. His comments are of a very different nature than a focus on the technology. It targets instead the "master" of the NASA and its proximity to political powers. "NASA's decline seems to have begun in 1971 with the administration of James Fletcher. A multimillionaire, he sits on the board of directors of many corporations that hold large contracts with NASA, and he has served in executive positions with some. He is from Utah, the home state of shuttle booster Senator Jake Garn, and the location of the important Morton Thiokol plant" (Perrow, 1986b, 354).

To support this, he cites a journalist of *The New York Times*, Stuart Diamond (1986a,b), who published a series of two articles revealing many problems with money spent as well as a number of audit reports indicating latent issues in the organisation of the agency. It is very obvious that his interpretation relies here on his tool view of organisations, which was described in a previous section, not on coupling and complexity.

Moreover, relying on such sources is consistent with his posture that "it is from the muckrakers, journalists, congressional committees, historians, and, occasionally, the economists and political scientists that we learn about the ways in which organizations shape our environment, not, ironically, from the sociologists" (Perrow, 1986a, 175).

But there is more about Challenger, if one, this time, introduces the issue of "organisational goal": "to be safe again, the space program will have to return to its original mission of scientific exploration, and separate itself from both commercial and military pressures" (Perrow, 1986b, 354). This is not quite a technological argument. The power of the elite, the political environment and the goals of NASA explain the accident. Not technology.

This interpretation, Perrow says in the 1990s, is "different from those of Starbuck and Milliken (1988a) and Vaughan (1990)" (Perrow, 1994a, 19). The difference is that his view is much more critical and targets explicitly failures of executives, confronting top management with the impact of their decisions on high-risk systems, something that is an enduring trait of Perrow's sociological stance and to be distinguished from the coupling/interactivity frame. Hence, his critics introduced previously about Vaughan's focus on cultural features of organisations at the expense of power.

A Neo-Weberian Approach to Disasters

A second convincing indication comes from a summary of *NA* in the third edition of his critical essay on organisational theory (Perrow, 1986a). In this summary, Perrow does not separate the technological from the organisational and social (and at times psychological and cognitive) dimensions of his analysis of major failures. Referring to his writings on organisations, he explains that he has "tried to make use of most the elements of the neo-Weberian model" (Perrow, 1986a, 147).

Of course, the coupling/interactivity framework is presented in this summary, but it is followed immediately by the macrosystem views offered about the airline and marine transport systems, the former qualified as "error-avoiding" and the latter as "error-inducing", as well as a reference to garbage can theory, which "highlights the unexpected interactions that can occur in reasonably complex systems" (Perrow, 1986a, 147).

To conclude on his contribution as a sociologist of organisations in the area of disaster and safety, he writes the following: "finally we took a specific problem, system failures, and examined the cognitive, structural, power and organizational environment aspects with our neo-Weberian model. The focus was on system interaction and control, whether at the level of the operator or elites deciding what kind of risks the rest of us should run" (Perrow, 1986a, 154). This last quote reveals the broader view that Perrow had about his work than what is usually stressed. The book is as much about technology as it is about organisations as conceptualised by Perrow.

This broader view will now be discussed more extensively, taking into account three categories: cognitive, organisational/managerial and macrosystemic. They correspond to different layers of explanation beyond the usual presentation of *NA*, which directly reflect Perrow's analytical lenses for studying organisations. Each of these categories is to a certain extent explored according to the industries that he investigates (nuclear, chemical, marine, etc.), depending on the availability of material. The two chapters dedicated to aviation and marine transport are the most extensive (probably because there was more information available about them at the time).

They contain a great level of information on several of the layers of issues that the three categories represent. Other chapters are not as elaborate, and the categories are not explored as deeply as they are in these two. Once combined, these three categories show a richer structure to the narrative of *NA* than the structural/technical one usually emphasised.

Task: A Cognitive Layer of Analysis

First, Perrow offers many insightful descriptions and interpretations of issues concerning human factors, related to automation and the mental constructs of operators facing dynamic work situations. This is an analysis of the interaction between technology and tasks; a core ingredient of the school of organisation theory is developed in the 1970s, as introduced earlier. Whether in aviation, marine transport or space missions, he applies an approach of "human error" that anticipates modern directions of cognitive principles applied in investigations and that relies on constructivist perspectives (e.g. Dekker, 2002). Thus, he offers convincing and alternative accounts of official investigative reports that interpreted accidents in rather simplistic ways when it comes to cognition.

His proposition that operators, pilots, astronauts or captains construct while performing their tasks mental images of situations that can at times be at odds with events is prescient. Of course, it is more likely to be so in complex systems where many hidden interactions can occur outside human understanding, but his examples are also proposed in other circumstances, as in marine transport where one encounters rather simple situations according to his criteria. Thus, the technological rationale is not entirely relevant. It is sometimes too late for these constructs to be adapted to operative constraints in order to avoid risky situations. Error, with these principles, is not the result of a "lack of something" (for instance, awareness) but a retrospective posture of an observer who can judge a situation and behaviour after the fact.

> We construct an expected world because we can't handle the complexity of the present one, and then process the information that fits the expected world, and find reasons to exclude the information that might contradict it. Unexpected or unlikely interactions are ignored when we make our construction. (...) in conclusion, I am arguing that constructing an expected world, while it begs many questions and leaves many things unexplained, at least challenges the easy explanations such as stupidity, inattention, risk taking, and inexperience.
>
> *Perrow (1999, 214)*

But he also produced a human factor engineering study from an organisational and sociological point of view, questioning the ability of high-risk systems to introduce these new approaches toward the design of safe automations and machines (Perrow, 1983). This study remains a reference today for those interested in understanding the problems faced when advocating human factor approaches in companies (Waterson, 2010).

Goal and Structure: An Organisational/Managerial Layer of Analysis

Second, his organisational and managerial appreciations of accidents are often rather subtle and ahead of many current ideas, even if, unsurprisingly, he very often formulates them with a critical tone. Although in many cases, he has a trenchant opinion for the organizational and managerial failures behind disasters, in many other places, he is rather cautious. This is consistent with his argument that some systems are more likely to suffer nas, as for instance in aviation, while also admitting that such systems are institutionally designed to be "error avoiding" (to be discussed in the next section). One initial and key concept for this sensitivity is the drawback of the retrospective posture. This has just been discussed when it came to operators dealing with risky situations in real time, but it is also a much more general statement that can be found in the first paper written about *NA*.

At the time, it was history that provided the analogy required to make this point. "Historians tell us the causes that should explain wars generally exist when no wars occur; in our passion to understand and create an orderly universe we explain the unusual event by invoking the usual and proclaiming it to be different, when of course it is not. Instead, it is the obscure, accidental, and even random concatenation of normal disorders that produces a great event that we assume must have had great causes" (Perrow, 1982, 176). This analogy was later formulated slightly differently to describe what would now be called a "hindsight bias" or "retrospective fallacy": "We say X and Y produced the accident, but if we find X and Y in many other plants but no accidents, where are we?" (Perrow, 1994a, 9).

With this in mind, he admits, even in systems with many resources dedicated to safety, such as aviation, that "with complex, tightly coupled systems with catastrophic potential, there is precious little room for management error, let alone other errors in the DEPOSE system" (Perrow, 1999, 141). But one explanation offered for management "error" is linked to the principle of "risk homeostasis", the principle which states that individuals, and by extension, organisations and systems, adjust their behaviour to change, so that they remain close to performance and safety borders.

This idea is expressed in the following sentence: "we continue to have accidents because aircraft and the airways still remain somewhat complex and tightly coupled, but also because those in charge continue to push the system to its limits" (Perrow, 1999, 123). Again here, in the end, the technological features are not really needed to provide a meaningful explanation of disasters from an organisational and managerial point of view. Systems can go beyond their limits while exploring the possibilities offered by the material and social resources.

This is reminiscent of Jens Rasmussen's idea, which has close connections to Perrow during the 1980s, when these ideas were exchanged and discussed, but it is also what has been recently stressed by Starbuck and Farjoun with the concept of "organisations at the limits" (Starbuck & Farjoun, 2005), which connects further investigations to the topic of goals as formulated by Perrow, nowadays described as strategy (more about this in Chapter 3).

For him, this perspective takes of course a judgmental twist: "unfortunately, the technological fixes have frequently only enabled those who run the commercial airlines, the general aviation community, and the military to run greater risks in search of increased performance. The technology improves, the increased safety potential is not fully realised because the demand for speed, altitude, manoeuvrability, and all-weather operations increases" (Perrow, 1999, 128). Similar comments are made for marine transport systems.

Environment: A Macrosystem Layer of Analysis

Finally, the last category derives from his broad sociological outlook about organisations' environment in the spirit of a society of organisations, which seeks to understand how interlocking systems made up of various organisations and their environments create global persistent patterns, sometimes detrimental externalities to society at large, as in the case of disasters. An example of this was presented earlier in his evaluation of the American healthcare system. This broad perspective is central to *NA*, and he extended this side of his contribution in a subsequent paper (Perrow, 1994a). This appears most explicitly when he compares aviation and marine transport systems.

The first is an error-avoiding system, whereas the second is an error-inducing one. Interpreted at an industry level, these two systems have indeed very different kinds of interactions between civil society (and associations), unions, professional, legal and state entities as well as private companies (including insurance). In this respect, in comparison to marine transport, "there is an enormous incentive to make commercial aviation safe" (Perrow, 1999, 127). Table 1.1 organises several quotes from Perrow that one can find about the macrosystem view of these two high-risk systems: aviation and marine transport.

What about the other high-risk systems studied in the book? Perrow offers comments throughout the book about the chemical and nuclear industry, for example, "The nuclear power industry, for example, lacks a strong union, has random public victims with delayed effects, has no safety board that is independent of licensing and regulatory functions, and does not see an immediate effect on its profits if safety lags (though a more severe incentive exists to avoid a catastrophic accident which could shut down the industry)" (Perrow, 1999, 127).

But later, in the 1990s, he offered to systematise this perspective and provided an interesting comparison of many high-risk systems on the basis of the material of the book and his macrosystem view (Perrow, 1994a), which was his way of contributing to the debate of how to characterise system safety, as with HRO studies (Grabowsky & Roberts, 1997).

Summary

After proposing this second reading of *NA*, what remains of technological determinism? Not much, although it is indispensable to consider technology, but not from a deterministic point of view. Cognition can be defeated because humans are intrinsically prone to construct fallible representation (task); organisations can be pushed to their limits when exploiting options to reach maximum performance at the expense (retrospectively) of safety (structure and goal), and macrosystem features that define specific

TABLE 1.1

A Macrosystem View of Safety in Aviation and the Marine Industries

	Aviation (Perrow, 1999, 127, 167, 168)	Marine (Perrow, 1999, 172, 173)
	Error-Avoiding System	**Error-Inducing System**
Civil society	Victims are neither hidden, random nor delayed, and can include influential members of the industry and Congress.	The identifiable victims are primarily low status, unorganized or poorly organized seamen. The third-party victims of pollution and toxic spills are anonymous and random, and the effects delayed. Elites do not sail on Liberian tankers.
Associations	A remarkable voluntary reporting system exists (Air Safety Reporting System).	The only international association concerned with safety is advisory and concerned primarily with nationalist economic goals.
Unions	There is a strong union at work to protest unsafe conditions – ALPA, the Air Line Pilots Association. It even conducts its own studies and makes its own safety recommendations.	Unions are weak.
Legal	Lawsuits follow immediately if the investigation of the NTSB hints at vendor or airline culpability.	The marine courts exist to establish legal liability and settle material claims, not to investigate the cause of accidents and compensate seamen.
State entities	The FAA is charged with both safety and facilitating air travel and air transport and spends significant amounts of tax dollars pursuing safety studies and regulations. An independent board, the NTSB, conducts investigations and prods the FAA to set new safety requirements.	The federal presence is minor and appears inept in the United States. It sets standards for those ships that want to use our ports, but the United States ranks 14th among nations in ship safety, so the standards cannot be very high. And finally, the only international association concerned with safety is advisory and concerned primarily with nationalist economic goals. Regulations succumb to economic and nationalistic pressures, and are highly ineffective.
Private companies	Airline travel drops after large accidents. Airline companies suffer if one of their models appears to have more than its share of accidents. Experience is extensive and the repetitive cycle of takeoffs, cruising, and landing promotes rapid training, precise experience with failures, and trial and errors for new deigns and conditions.	Shippers do not avoid risky "bottoms" but pick the cheapest and most convenient, and cannot choose to stop shipping for a time because the last cargo was lost. The insurance company is a passive contributor, passing on the costs to the final consumer.

FAA, Federal Aviation Administration; NTSB, National Transportation Safety Board.

institutional configurations generate different degrees of likelihood and frequency of accidents and catastrophes (environment).

None of these concepts need to be specified concerning technology; they appear rather independent of it. However, in any empirical study, technology cannot be left aside when trying to understand precisely how individuals interact with it. But these concepts are part of the book and contribute to a rather coherent whole, providing principles to interpret accidents.

This is certainly one of Perrow's greatest legacies to the field. These concepts add tremendously to our knowledge of how safety or, conversely, how disasters are produced. Instead of a monodimensional vision based on technology, the narratives available in *NA* offer a multidimensional view and several layers of analysis that help to sensitise to the numerous stories that Perrow gathered (Table 1.2).

This is, in the end, not surprising when one considers Perrow's wider sociological contribution found in earlier and subsequent books (briefly summarised in this chapter in the first section), a contribution that goes well beyond the rather specialised topic of industrial safety. For instance, his interest in technology is a key feature of his approach to organisations more generally, an approach that had a very central role in relation to organisations' activities and structures, in accordance with the "contingency" school of the 1960s and 1970s.

This school contained a deterministic slant, not unlike *NA*, which he had to challenge and return to in later years "something called 'contingency theory' today, to which I have contributed, would be shocked to learn that technology dictated no particular structure or leadership style" (Perrow, 2002, 174). His strong emphasis on both technology and operators in relation to disasters indicates the underlying influence found in *NA*. His interest in the goals and environment of organisations is also strongly reflected in *NA*, when he comments on how organisations can be pushed beyond their limits by their masters as much as when he describes how the environment shapes the likelihood of catastrophes.

TABLE 1.2

The Multidimensional Approach (and Legacy) Found of *Normal Accident*, beyond Technological Determinism

Technology	**Cognition (task)**
Proneness of accidents is a product of coupling/ interactivity of high-risk systems	Humans tend to naturally construct mental models of the world, in an increasingly automated working environment
Organisation/managerial (structure and goal)	**Macrosystem (environment)**
Organisations can be pushed to their limits and can reveal in certain cases gross negligence of failing executives	Proneness to accidents is a product of specific configurations between civil society, legal and state entities, unions, private companies and associations

Implications of the Second Thesis

What Does "Normal" Mean?

Now, to return to the point of departure of this chapter concerning the repetition of accidents in high-risk systems, it is interesting, based on the above developments, to raise the question of the na thesis and Perrow's insights on this specific issue. First, it is interesting to notice that Perrow had another expression for naming the kind of phenomena that he was trying to conceptualise. He proposed "system accident" as a synonym to normal accident. But it is normal accident, the title of the book, which has been retained.

This is not without ambiguities, because the word "normal" has had different meanings over the years in this field of research. First, "normal" was used to differentiate investigations following accidents from studies studying daily operations (such as the HRO research tradition). This second notion of normal conveyed a very different meaning than Perrow's initial one. This alternative meaning suggests that there are some usual ways of doing things or of performing activities that one can expect to observe in real-life situations.

One problem is of course to characterise these usual ways. Secondly, normal has been retained by those who extended Perrow's initial statement to argue that accidents are unavoidable due to techno organisational reasons and not only technological ones (Vaughan, 1996). This is a third meaning. A final meaning could be suggested now, which uses the notion of normal from a historical, empirical but also normative perspective to state that accidents and catastrophes are normal because their repetition across critical safety systems makes them so (empirical), despite knowing that they could be prevented in principle by adequate managerial and regulatory practices (normative). Table 1.3 puts together these four different meanings of normal in relation to safety and accidents.

TABLE 1.3

Several Meanings for the Word "Normal" in Safety and Disaster Research

	Meaning
Normal (system) accident (1)	Hidden interactions create unpredictable events (technological determinism)
Normal operations (2)	In opposition to accident investigation, normal operation is about what is observed and/or expected in daily operations of high-risk systems
Normal accident (3)	Extended version of (1), accidents are inevitable products of technical/organisational/social dynamics (sociotechnological type of determinism)
Normal accident (4)	Different version of (1) and (2) and critical version of (3): it is the repetition of (component) accidents in the past 30 years that makes them now normal, despite knowing that they could be prevented in principle

In fact, Perrow used the term "normal" in several different ways in *NA*. Some of them were very close to the variations shown in Table 1.3. I found at least two examples. In the chapter on marine transport and its error-inducing system property, he writes "There are so many source of failure, they are encountered continuously, often as a matter of course, and become indistinguishable from normal operations much of the time" (Perrow, 1999, 176).

The second is in his appraisal of the situation following the Flixborough accident, in the chemical industry. "There was organizational ineptitude: they were knowingly short of engineering talent, and the chief engineer had left; there was a hasty decision on the bypass, a failure to get expert advice, and most probably, strong production pressures. But as was noted in the case of TMI and other nuclear plants, and as will be apparent from other chapters, this is the normal condition for organizations; we should congratulate ourselves when they manage to run close to expectations"(Perrow, 1999, 111).

This pessimistic view is slightly different than the technological determinism of the *NA* thesis. It combines meanings (2) and (3). Accidents may be normal, in certain circumstances, because organisations can't be expected to behave as we would like them to. Therefore, "we should not expect too much of organizations" (Perrow, 2011b, 9). When considering these different uses of the word in the book, Perrow seems very close to the extended versions of *NA* of Diane Vaughan and Scott Snook. But he has rejected the former's version ("I disagree with her interpretation," Perrow, 1999, 379) and endorsed the latter's ("Snook makes another contribution. This is, he says, a normal accident without failures," Perrow, 1999, 78). What conclusions can we draw?

Does the Book *Normal Accident* Make Any Sense?

If we come back to all of what has been said about Perrow and *NA*, a certain number of observations have been made, which contribute to a significant deconstruction of the traditional presentation of the first thesis of Perrow's book *NA*:

1. The technological orientation of the main thesis of the book is very strong but is not the only possible reading.
2. There is a much richer quality in the narrative that contains many more conceptual inputs worth extracting because they are very consistent and show that Perrow goes well beyond a focus on technology by considering cognition (task), organisation (structure, goal) and macrosystem (environment) views (Table 1.2).
3. This creates many tensions in the book because the structural features (coupling/interactivity) of the central thesis do not account for the diversity of cases and the fact that for Perrow, accidents are sometimes the result of gross negligence and/or organisation/executive failures (which he considers partly to be expected, namely "normal," which introduces a different meaning to the word, Table 1.3).

4. Perrow not only makes this quite explicit by rejecting what he considers the misuse of his technological argument by other authors for explaining disasters but also criticises what he sees as inappropriate extension of his idea of na, as for instance with Vaughan's interpretation of the Challenger accident.
5. However, he also endorses this type of extension to his initial central thesis when, for instance, Snook's interpretation of friendly fire shows the incremental decoupling of parts in a complex civil/military system through practical drift.

How can these various theses be integrated into a meaningful argument? Clearly, the situation is now quite complex. Nevertheless, I think that it makes sense if one comes back to the tensions introduced earlier and to Perrow's intellectual posture as well as the problem of our appreciation of accidents more generally. It is now time to consider the following question: "Was Perrow right for the wrong reasons?"

Was Perrow Right for the Wrong Reasons?

One interpretation is that Perrow is the first to really attempt a comparison of high-risk systems from a sociotechnological perspective that is at the same time critical and technologically based considering the nature of these systems (derived from his experience of TMI). In so doing, he had to bring together a very wide range of dimensions, in accordance with his background in sociology and with only few available studies in this area of investigation.

He had to innovate and therefore relied on his sociological background, and these dimensions exceeded his technological rationale. Analyses at the cognitive (task), organisational/managerial (structure and goal), and macrosystem (environment) view (Table 1.2) were developed to support the interpretations of the diversity of cases provided by the reports, studies and articles available.

Despite this intrinsic difficulty and the tensions that were created through this strategy (as discussed earlier), he maintained his initial thesis based on a strong technological argument and wrote the book with this in mind, maybe as a way of challenging the nuclear industry if one considers his conclusions (abandoning nuclear power plants).

But, when, in the 1980s, high-profile accidents and catastrophes occurred, what originally exceeded the technological rationale of *NA* came to be the most relevant concepts for interpreting them. These, such as the Challenger and Chernobyl in the 1980s, were not products of tight coupling and complexity but the results, through a tool view, of organisational and executive failures, a view quite consistent for a critical sociologist, as developed by Perrow in many of his books that were not concerned with safety.

And, when Perrow endorses Snook's extension of na from a sociotechnological angle, it is because the organisations involved had very little in common with private ones, in which production pressures, gross negligence and failures are, according to him, to be expected. Military and (nonprofit) civil organisations interacting in the background of Snook's case study, in Perrow's view, are not related to for-profit organisations and their ability to concentrate economic and political power at the expense of safety, by, for example, limiting regulatory oversights.

A New Version of Normal Accident?

With this is mind, one therefore understands quite well his reactions following the Fukushima Daïchi disaster in 2011. "Nothing is perfect, no matter how hard people try to make things work, and in the industrial arena there will always be failures of design, components, or procedures. There will always be operator errors and unexpected environmental conditions. Because of the inevitability of these failures and because there are often economic incentives for business not to try very hard to play it safe, government regulates risky systems in an attempt to make them less so" (Perrow, 2011b, 44).

But when writing this, one understands also fairly well that the meaning of normal has slowly evolved from a technological sense as initially conceived to a more organisational, managerial, social and political one. It is, as a consequence, quite tempting to conclude that Perrow was right, accidents are normal, but for the wrong reason. They are not the product of technological determinism.

How can we formulate this new version of na to move beyond the initial technological emphasis? I believe that the fourth meaning of na suggested in Table 1.3 captures best the second thesis of *NA* (Table 1.3). "Accidents and catastrophes are normal because their repetition across critical safety systems makes them so, despite knowing that they could be prevented in principle, by adequate managerial and regulatory practices".

Some Limits of Perrow's Argument

But this second reading leads us to identify some limitations. Some are worth pondering. Three are suggested here. One issue is the exclusive reliance on accident reports. The main problem is that it is difficult to assess situations in comparison to what might be expected. To admit that organisations have been pushed to their limits is a now obvious interpretation when considering accidents retrospectively. When can we be sure that a system remains within acceptable boundaries when everyone accepts that nothing is perfect and that the line is difficult to draw? It is certainly looked at from a critical angle for Perrow, who often, even without any in-depth data analysis, seems to be inclined to favour not only the option of criticising management and executives for their failures but also regulatory oversights.

It is indeed appealing to do so and, in many cases, probably legitimate, but empirical data are needed to establish conclusive interpretations of this sort. This is something that Perrow is aware of. "Classifying something as an executive failure rather than a mistaken executive strategy or a poorly performing executive, or even a failure by management or workers, is controversial. Observers can disagree, and since it is easy to have a mistaken strategy or a poorly performing organization, it takes a great deal of evidence to make the case for executive failure" (Perrow, 2011b, 293).

Thus, a second issue is, given this critical posture, that using a tool view of organisations can tempt analysts to make quick conclusions about executive failures. The same applies to regulatory failures. In the light of any catastrophe, the concept of "regulatory capture" is very often the core category that Perrow uses, but as political scientists have also warned "observers are quick to use capture as the explanation for almost any regulatory problem, making large-scale inferences about agencies and their cultures without a careful look at the evidence" (Carpenter & Moss, 2014, 5).

A last limit to be introduced here is the absence in Perrow's work of any criteria that would help assess companies individually. Comparisons at a macrosystemic level do not help discriminate between "single" organisations. This is obviously a problem because for each industry, regardless of its proneness for error, some organisations will experience a major accident and some will never, despite sharing the wide configuration of an industry.

This, of course, has been explored in research traditions with topics such as "safety culture" or "high-reliability organisations", but there is a need to include more Perrow's indications about executive strategies in their environment (e.g. legal, industrial, financial) as well as their relation with regulatory oversight to better understand how limits are crossed (Chapter 3 explores this issue).

A good contemporary example, yet again retrospective, is British Petroleum (BP) and its series of accidents in the United States between 2005 and 2010, in relation to masters' strategies and a specific regulatory environment and discussed in Chapters 3 and 4 (Bergin, 2012). One problem is that Perrow never goes beyond this implied critical view and second thesis. To explore further the implications of this second thesis, one needs to turn to Hopkins, something done in the next chapter.

Summary of Chapter 1

Starting with the traditional view that the book *NA* is based on a deterministic argument, it is shown that there is more to it. In order to do so, the classic criticisms are introduced and challenged because they often consider the book from a narrow perspective, leaving aside Perrow's broader sociological contribution. In contrast, it is shown that there is much to gain from positioning *NA* within the author's wider intellectual trajectory. Beyond writing about disasters, Perrow is a sociologist with a critical, power, material and macroapproach to organisations. It is this background that one finds in the rich narrative of *NA*.

Perrow is a leading author in the field of sociology of organisations which he understands through analytical categories combining technology (and tasks), structure, goal and environment of organisations. Moreover, he grants organisations a central status in our description and understanding of societies. Because of their prominence through wage dependency, factory bureaucracy and the externalities, it is vital to fully embrace the extent with which societies are shaped (but also shape in return) organisations, product of a specific relationships between states and for-profit ones.

It is with this analytical background in sociology that he advocates, beyond the coupling/interactivity framework, an analysis of cognitive (e.g. human actively constructed mental images of their environment in the context of their tasks), organisational/managerial (systems can exceed their limits due to goal orientations and choice of structure by leaders) but also macrosystemic (configurations differ between industries characterising a specific environment more or less conducive to safety) aspects of accidents.

Leaving behind a presentation of *NA* that is monodimensional, one sees instead a much more multidimensional interpretation, the second thesis of *NA* the book, and the possibility of a new formulation of na, beyond the first original, technologically centred version (and its extended sociotechnological one). This second thesis is the critical one. "Accidents and catastrophes are normal because their repetition across critical safety systems makes them so, despite knowing that they could be prevented in principle". This chapter is therefore a first step from *NA* to *Post Normal Accident*. To move to the next one, one now needs, as done for Perrow, to read more closely the work of another sociologist, Andrew Hopkins.

2

Hopkins, the Unofficial Theorist of NA

Introduction

Hopkins is a successful safety writer. In the past 20 years, in the first two decades of the 21st century, he has become renowned for his analyses of several technological disasters and sold tens of thousands of his books to a wide audience of practitioners. When *Normal Accidents* (*NA*) was released in 1984, his contribution to the field of accident, safety and technological risks was marginal, with two or three published articles or book chapters, quite unknown to mainstream safety research in the United States and Europe. At the time, Hopkins was a sociologist in Australia with a critical, white-collar crime (WCC) and sociolegal mindset and was unaware of Perrow's *NA*.

It is only in the late 1990s that Hopkins fully engages first with disaster analysis and second with mainstream sociology of safety, including Perrow. As commented in Chapter 1, Hopkins rejected *NA*. His own analyses of accidents convinced him about the limits of the normal accident (na) thesis. Accidents are not the products of complexity and coupling, but the products of lack of sufficient concern for safety by top management of organisations (and regulations). This, in fact, proves to be quite consistent with the second thesis of *NA*, as developed in Chapter 1, namely the idea that *accidents are normal because they repeat despite knowing that they could be prevented in principle.*

But Hopkins does not stop where Perrow stops. Despite sharing with Perrow a critical view of the powerful, Hopkins goes one step further by performing several times in-depth analyses of single cases of accidents in the mining and petrochemical industries. And, contrary to the dominant interpretation that he is a successful storyteller, I argue that he is also a theory builder. I go one extra step and contend in this chapter that he is the unofficial theorist of the second thesis of *NA*. When Perrow stopped at an implicit level of formulation that I had to make explicit in Chapter 1, Hopkins produced the material to ground the second thesis. This, however, needs to be extracted from his writings because it has been overlooked so far. The aim of this chapter is to justify and clarify this assertion.

First, this chapter introduces Hopkins' analysis of disasters with the help of one of his visualisations, the causal diagram of the Longford explosion in Australia in 1998. Second, I show that Hopkins' success attributed to his storytelling skills must be unpacked to reveal the persuasive narrative sequence repeated across the themes of his books. It is this narrative structure that provides the ingredients attracting practitioners' interests and convincingly manages to reinforce the logic of the second thesis of *NA*. Third, I argue that one needs to complete the analysis of Hopkins' work by revealing the normative model from which his storytelling unfolds, and fourth, that he is in this respect the unofficial theorist of the second thesis of *NA*.

Hopkins, the Storyteller

Between 1999 and 2012, Hopkins released four in-depth analyses of disasters: the Moura mine explosion in 1994 in Australia (Hopkins, 1999a), the gas station Esso explosion in 1998 in Longford, Australia (Hopkins, 2000), the BP Texas city refinery explosion in the United States in 2005 (Hopkins, 2008) and the Macondo well, Deepwater Horizon explosion in the Gulf of Mexico in 2010 (Hopkins, 2012), plus one study in collaboration with Jan Hayes on two gas pipelines disasters in the United States in the early 2010s (Hayes & Hopkins, 2014). Three additional but shorter accounts of other events (in railways, air force defence and mining) are also available in books published in the first decade of the 21st century (Hopkins, 2005, 2007a).

Studying and Visualising Accidents

In all of these disaster books, one follows a trail going (metaphorically) from the bottom to the top, starting with technical breakdowns and human errors to organisational issues including risk analysis, learning from incidents, organisational structures, incentives and rewards, finishing with top executive leadership and regulation of safety (more about this below). Served by appealing visualisations in several accounts applying the AcciMap layout of Rasmussen (Rasmussen, 1997), Hopkins shows convincingly that one would fall short of an explanation for disaster if one stopped at the technical or human error level of description (Figure 2.1).

What Do We See in This Picture?

First, we see five levels of analysis and causation, from bottom to top: (1) physical accident sequence, (2) organisation, (3) company, (4) government/regulatory system and (5) societal. The map selected earlier is applied to the explosion of an Esso gas station in Longford Australia, causing the death of

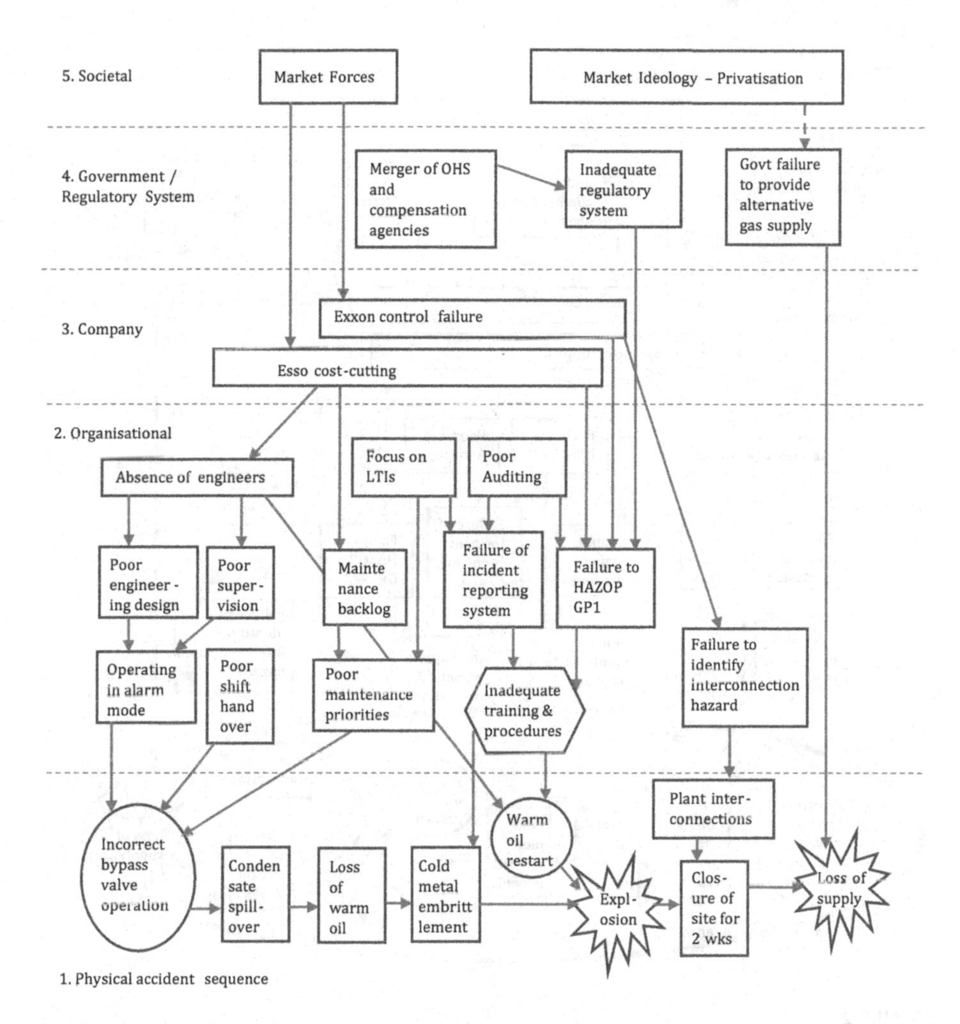

FIGURE 2.1
An AcciMap of the Longford disaster by Hopkins (Hopkins, 2000).

several workers and stopping households' gas supply for several weeks. At the bottom is thus the physical accident sequence. This is where Hopkins' account starts (Figure 2.2).

As one can read it, concentrating for now on the bottom left part of the representation, that an incorrect bypass valve operation causes a series of physical and engineering consequences: the contact of released warm oil on a brittle cold metal (brittleness due to improper handling by operators of a sequence of operations, the bypass) leading to an explosion. One classic interpretation is to reduce the event to what triggered it initially, namely an incorrect valve operation. An aggravating problem is the interconnection of two plants, causing the long period of time without supply because, instead

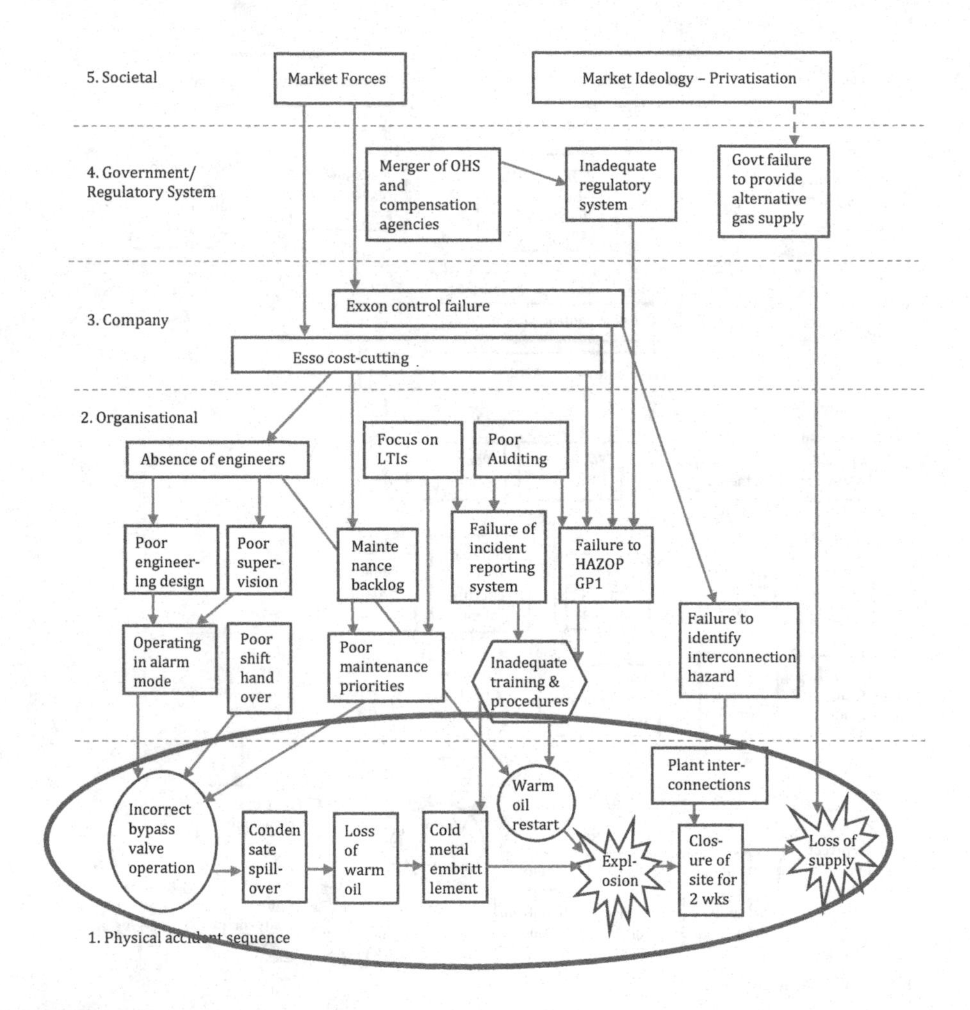

FIGURE 2.2
Physical accident sequence in the Longford disaster.

of affecting just one plant where the explosion occurred, it affected two plants. The bypass operation and the interconnection are Hopkins' points of departure of his analysis.

From there, he considers organisational issues (Figure 2.3). The incorrect bypass operation is due to *poor engineering design* of the alarm system and *lack of supervision* that both explain why alarms are actually bypassed daily (not only the day of the accident), without leading to any corrective actions by management. Combined with poor maintenance practices, fieldworkers were, in fact, prior to this event and for a certain amount of time, running operations in a degraded mode without corrective actions by management, although the situation could have easily been known.

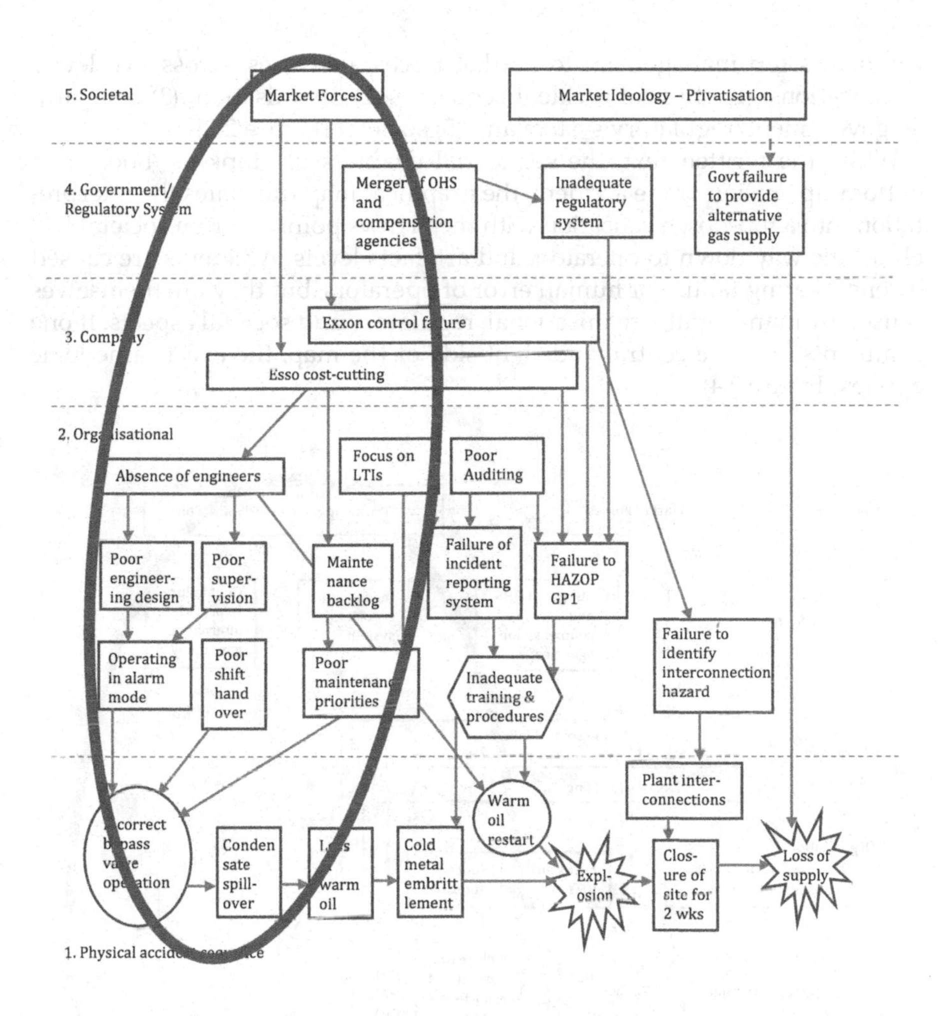

FIGURE 2.3
Causation from market forces to incorrect bypass through cost cutting.

We now go up one notch in the causal diagram in Figure 2.3, exploring further the content of the organisational level and reaching higher levels, company, regulatory and societal ones. The *absence of engineers* on site, working away from operations in an office at corporate headquarters, contributed to the absence of reaction from management regarding workers' practices and of available expertise on site which could have helped improve design, maintenance and learning.

And, the decision to remove the site engineer is directly linked in Hopkins' diagram to *Esso's cost-cutting* orientation, a company-level description, due to *market forces*, a societal level of analysis. Therefore, from bottom to top, from incorrect bypass operation by operators through decision-making of company's top management to market forces, one goes across five levels of causations: (1) physical accident sequence, (2) organisation, (3) company, (4) government/regulatory system and (5) societal (Figure 2.3).

While the written text, the sequential chapters of Hopkins' book, is a bottom-up version of the accident, the graphical map translates the interpretation into a top-down causation with the arrows pointing from societal levels all the way down to operators and artefacts levels. Accidents are caused by engineering failure or human error of operators, but they are themselves caused by managerial, organisational, regulatory and societal aspects. If one comments now the central and right side of the map, the exact same logic applies (Figure 2.4).

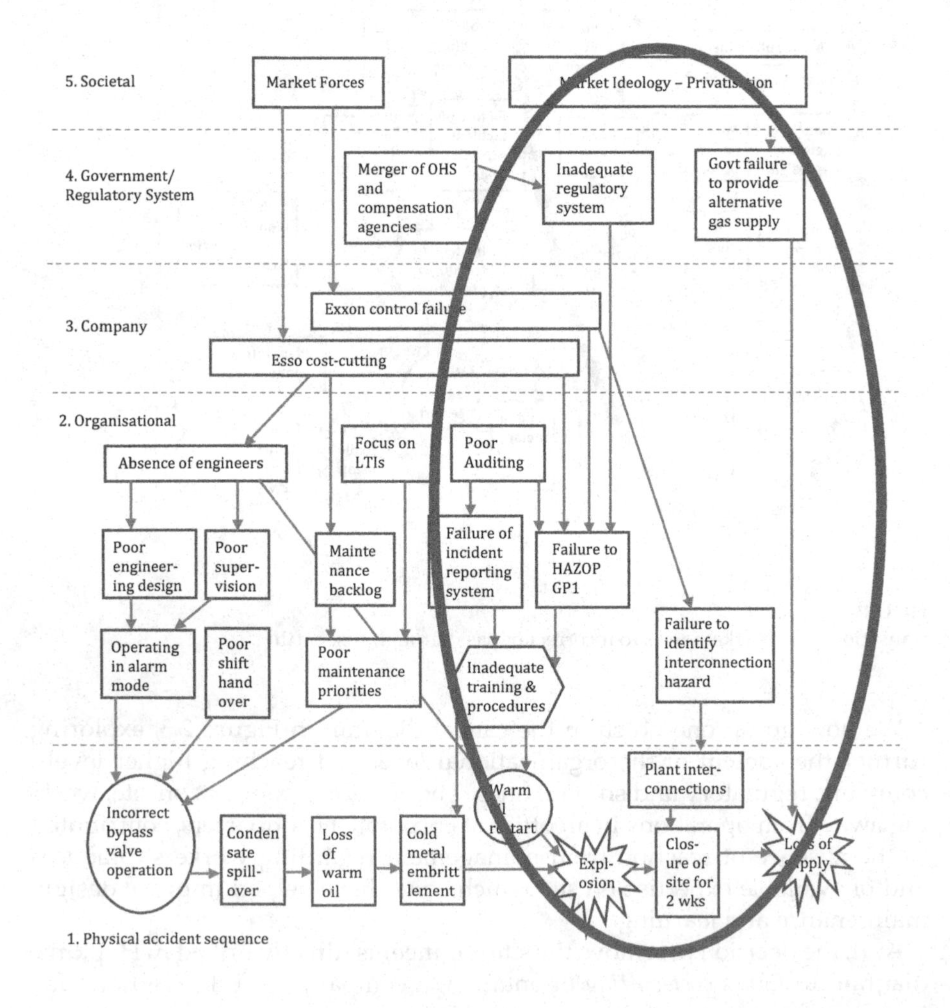

FIGURE 2.4
Causation from market privatisation ideology to explosion.

Incorrect bypass valve operation, warm oil restart and plants interconnections are linked to other key organisational features including risk analysis (*Hazop failure*), learning from experience (*failure of incident reporting system*), indicators (*focus on lost time injuries*) and auditing (*poor auditing*) at the organisational level of analysis. But these are also in turn then connected to the company level of explanation with *Esso cost cutting* and *Esso control failure* and then government/regulatory system with the two items of *inadequate regulatory system* and *government failure to provide gas supply*, themselves linked, another level above, to the societal level, with *market ideology of privatisation*.

So ultimately, one can read in one causal diagram that market forces and ideology of privatisation penetrated and translated into Esso's decision-making processes to create the favourable organisational conditions (cost cutting, lack of control) for human errors and technical breakdown to occur the way they did. This visualisation therefore greatly supports Hopkins' book chapters by connecting them together in one articulated graphic made of short texts, boxes and vertical arrows crossing horizontal lines representing levels.

Visualisations of Hopkins and Perrow

Visualisations play a key role in safety research (Le Coze, 2019c), and it is interesting for this reason to contrast this graphic with Perrow's matrix, reproduced in Chapter 1 (Figure 2.5). These two pictures exemplify the point made in Chapter 1 and the point currently developed in this chapter. On the one hand, Perrow developed a framework (the matrix, Figure 1.1) in which he could comment his distinctions between normal and component accidents but without focusing in depth on a single case of disaster.

One value of *NA* is the identification and then its categorisation of high-risk systems, which is visualised in this matrix of coupling and interactions. High-risk systems situated at the top right corner are likely to experience na because of tight coupling and complexity, whereas in other quadrants, with low coupling and simple interactions, accidents are not normal ones, but component ones (according to Perrow).

On the other hand, Hopkins concentrates on a single event at a time. One value of Hopkins' accounts of accidents is the systematic (i.e. many themes are explored and discussed) and systemic (i.e. multilevel and multidimensional) analysis, which is represented by his causal diagram (Figure 2.1). By probing deeper than Perrow into each event, he first comes to reject the argument of Perrow about the normality of accident (na) because each time it seemed reasonably possible to prevent disasters (Hopkins, 1999b), but, second, he also comes to nuance, over the years, the critical tone of his analysis (Hopkins, 2016), a tone that Perrow has always maintained (Perrow, 2015). This last point is commented more, later in this chapter.

However, looking at the causal diagram, one might say about Hopkins that his message is not really new in the 2000s when he starts publishing his case studies. Turner (Turner, 1978) and Perrow (Perrow, 1984) in the 1970s and 1980s

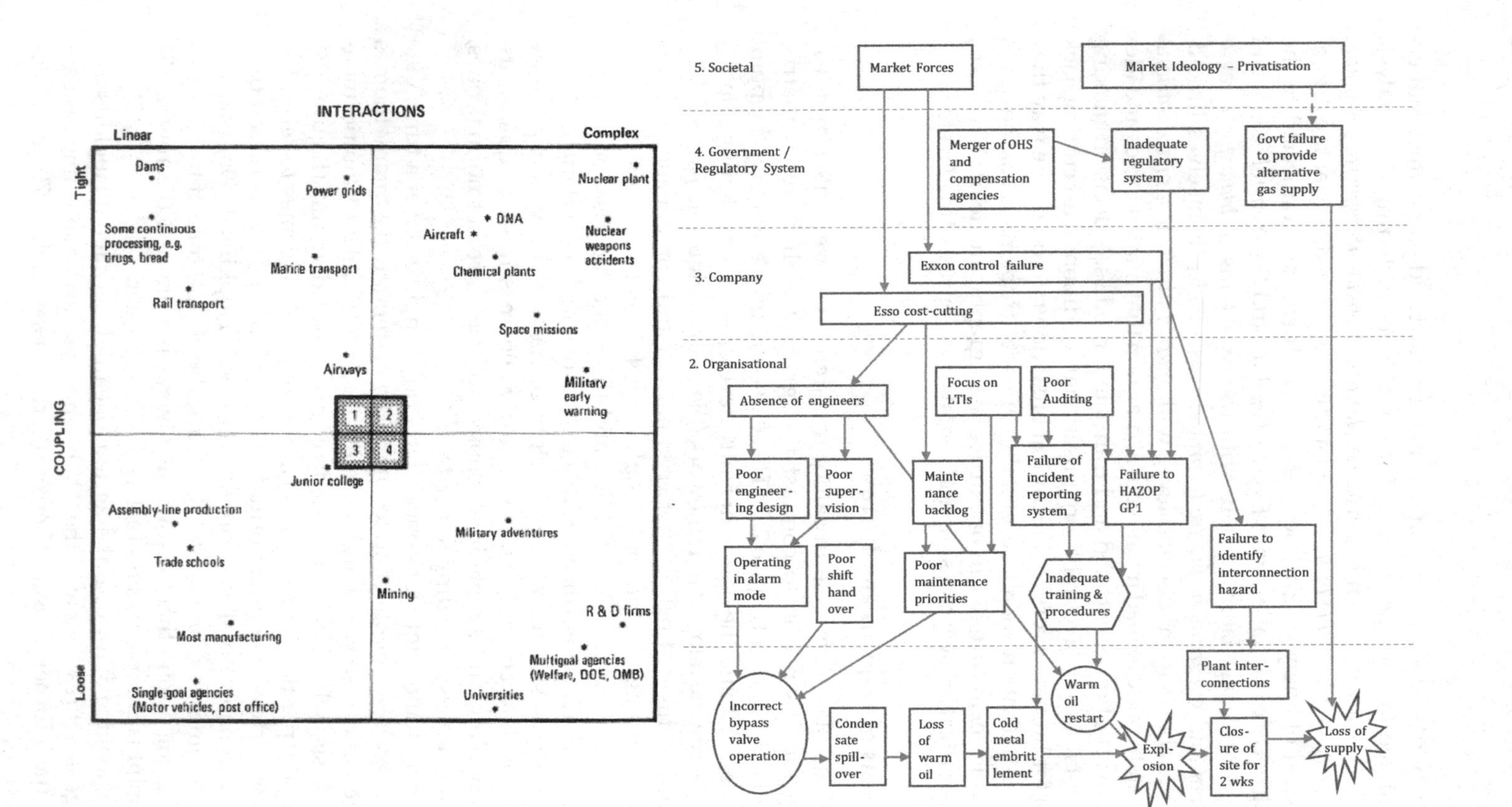

FIGURE 2.5
Comparison of Perrow's and Hopkins' visualisations.

already promoted such an idea. Plus, Reason popularised the organisational accident view in the 1990s with appealing models (Reason, 1997), whereas Vaughan did it from an ethnographic point of view with a 10 years' study of the National Aeronautics and Space Administration (NASA) (Vaughan, 1996).

But the success of Hopkins' books is high (Hopkins, 2016). This achievement is a credit to the engaging writing style of Hopkins, his ability to appeal to an audience of both practitioners and scholars interested in safety. Another facet of this success is that he has been writing books for an industry, the petrochemical industry, which has been behind for quite some time in comparison to other high-risk industries such as aviation and nuclear industry in the importation of knowledge developed by social sciences for safety purposes. Hopkins has played with his books an important role of translator in this respect for an industry that was not familiar with sociology.

Returning to the difference between Perrow and Hopkins, Hopkins probing deeper into single cases, one explanation is that Perrow wrote about disasters as a topic among other more broader ones (i.e. organisation theory, society of organisations), whereas Hopkins focused on safety and accident for 30 years after an early start in WCC (Hopkins, 1978a,b), contributing to the development of a sociology of safety with a normative side appealing to industry and regulators as attested by the size of his readership among practitioners.

Generally, this popularity is attributed to his storytelling skills although this ability is not really defined, and only alluded. "Andrew Hopkins is a consummate storyteller as well as being an internationally known expert on the breakdown of hazardous socio-technical systems. I believe that only stories such as those told here can capture the subtle influences of organisational culture and embrace the complex interactions between causes and conditions" (Reason, 2005, vi). Let's elaborate next on Hopkins' storytelling.

Writing Successful Stories

A Narrative Structure

If one probes deeper into these storytelling qualities, one finds a recurring pattern or a structure of narrative based on sequences associating descriptions (1), assumptions (2), explanations (3), comparisons (4), recommendations (5) and counterfactual reasoning (6). It is precisely, I contend, because Hopkins articulates several of these sequences that he is a popular safety writer.

The first sequence, description, is about bringing facts about what happened. It targets one aspect of the multidimensional nature of an accident. The second sequence, assumption, characterises a taken-for-granted way in the industry of dealing with the topic described. Explanation brings social sciences insights into a particular phenomenon, whereas comparison, the fourth sequence, identifies a similar accident, relevant academic models or useful practices in industry to open the case to wider horizons.

The fifth and the sixth sequences close the argument by proposing solutions to deal with the problem (recommendation) and assert that these solutions would have helped decrease the likelihood of the event (counterfactual reasoning). This last sequence is turned into a typical formulation "had this been done this way, the accident would have been avoided".

Hopkins is of course aware of this sort of combination of analyses and recommendations in his accounts. Commenting on his first disaster book, he writes that "The book was an examination of these failures coupled with suggestions for correcting them. Most of my subsequent books have followed this model" (Hopkins, 2016, 34). However, it proves interesting to go beyond this first level of characterisation with a fuller unpacked narrative structure.

Note before illustrating that this narrative structure remains a characterisation of an ideal that is not applied thoroughly as such across the themes put together in his disasters books, or in this specific order, but functions as the engine behind the acclaimed Hopkins storytelling. Let's now illustrate with one theme extracted from the case introduced earlier (Figure 2.1), auditing.

An Example of Narrative Structure: Auditing

In Figure 2.1, the box "poor auditing" at the organisational level of analysis is linked, with the help of two arrows, to "failure in incident reporting system" and "failure to Hazop G1". These two links mean that because auditing was not good enough, it missed the opportunity to improve risk analysis and learning practice for the safety management of the site of the company. It is therefore an important activity.

In the text, auditing is discussed in a chapter dedicated exclusively to this topic (Hopkins, 2000, Chapter 7). In this chapter but also elsewhere in the book, one can locate the sequences of the narrative structure defined earlier, articulated to produce the storytelling effect which is behind, I contend, the success of Hopkins' books. I extracted quote from this chapter and from other parts of the text to show the existence of such articulated sequences.

First, Hopkins describes the problem: "It is clear that there was plenty of bad news around (…). But the Exxon audit missed it all" (2000, 93). This is the sequence description (1). This quote introduces what the rest of the chapter unravels. Why, if there is sufficient information in the organisation about many safety concerns, no action has been taken before the disaster despite audits carried out?

Then, an assumption (2) of the organisation and associated with auditing is introduced. This assumption is extracted from the testimony of a managing director. "Six months prior to the explosions, Esso's health and safety management system (called OIMS – Operational Integrity Management System) was audited by a team from Esso's corporate owner, Exxon (…) Esso's managing director reported to the inquiry that the audit had shown that most elements of the safety management system were functioning at level three or better" (Hopkins, 2000, 81).

This additional sequence contrasts now, on the one hand, the presence of many bad news which are available when interviewing people retrospectively, and, on the other hand, positive auditing results. This assumption considers that a good principle of auditing is to rank sites with score card (from 1 to 5) based on predefined level of compliance to a defined standard, but one now understands that it might not be appropriate.

What is interesting here is that this narrative step can echo particularly well practitioners' experiences beyond the specific organisation involved in this event. Many in the industry use this type of auditing technique, and they are therefore quite interested in finding out about its limitations. This is what the explanation (3) sequence deciphers, as captured in the following selected quote.

"It is worth pointing out that an audit whose purpose is to identify hazards which have been missed does not lend itself to this score card approach" (Hopkins, 2000, 86). Indeed, in this accident, a risk analysis was not carried out to identify the accident sequence which created the disaster. So, only an engineering approach that would have scrutinised which scenario was missing would have been likely to identify gaps, not a score card one based on standard compliance.

So, an explanation is now available. There was bad news, but auditing failed to collect them because it was not designed to do so. Is this new? Are there other similar cases? The comparison (4) sequence shows that it is not new. Moreover, it shows that such accidents can be prevented by applying better safety management principles developed elsewhere or by following properties of organisational reliability observed by research in other contexts.

So, it is far from new and "one of the central conclusions of most disaster inquiries is that the auditing of safety management systems was defective (…)" (Hopkins, 2000, 54). Piper Alpha is an example. "Following the fire on the Piper Alpha oil platform in the North Sea in 1987 in which 167 men died, the official inquiry found numerous defects in the safety management system which had not been picked up in company auditing" (Hopkins, 1999b, 29).

The complementary option is to compare the case with known good practices in other industries or companies, and Hopkins refers to the experience of a mining corporation. "BHP coal had clearly learnt the lesson of its previous auditing failure. It had understood that the hallmark of a good audit is that it must be thorough enough to uncover the bad news about safety and convey it upwards to the top of the corporation" (Hopkins, 1999b, 35).

If this was not convincing enough, safety research also insists that it is precisely "the strategy which HROs adopt is collective mindfulness. The essence of this idea is that no system can guarantee safety once and for all. Rather, it is necessary for the organisation to cultivate a state of continuous mindfulness of the possibility of disaster "(Hopkins, 2000, 140). But, in the case of Esso, instead, "safety auditing, an ideal opportunity to focus on the possibility of failure, was turned into an opportunity to celebrate success" (Hopkins, 2000, 142).

The previous sequences, description (1), assumption (2), explanation (3) and comparison (4), prepare the reader for the next two sequences, recommendations (5) and counterfactual reasoning (6). Indeed, once the reasons for poor auditing have been addressed, one can suggest to correct them, with the help, for instance, of "a rigorous audit needs to examine the hazard identification strategy and make some effort to seek out hazards which may have been missed" (Hopkins, 2000, 86).

The efficiency of such a solution can now been tested hypothetically through a counterfactual argument. "Had it reached the highest levels of the company and had been acted on, it would have averted the incident" (Hopkins, 2000, 93). In other words, had auditing been designed to operate as suggested to be more focused on gaps in the identification of technological risks, things would certainly have been different, and the accident prevented.

Box 2.1 compiles the sequences for the summarised narrative structure applied to "poor auditing", a narrative structure that could have been illustrated for other themes developed in different chapters, and indicated on the causal diagram (Figure 2). The point is that, by combining these different rhetorical elements together within a broader perspective of the case, Hopkins does bring readers not only an organisational explanation but also possible solutions to these organisational problems. It has an engineering side to it, what could be called a socioengineering side.

BOX 2.1 ILLUSTRATION OF THE NARRATIVE STRUCTURE APPLIED TO AUDITING

Description (Retrospective) (1)

It is clear that there was plenty of bad news around (…). But the Exxon audit missed it all.

Hopkins (2000, 93)

Assumption (Taken for Granted In the Case Study) (2)

Six months prior to the explosions, Esso's health and safety management system (called OIMS – Operational Integrity Management System) was audited by a team from Esso's corporate owner, Exxon (…) Esso's managing director reported to the inquiry that the audit had shown that most elements of the safety management system were functioning at level three or better.

Hopkins (2000, 81)

Explanation (3)

It is worth pointing out that an audit whose purpose is to identify hazards which have been missed does not lend itself to this score card approach.

Hopkins (2000, 86)

> ...and as a consequence, one of the central conclusions of most disaster inquiries is that the auditing of safety management systems was defective.
>
> **Hopkins (2000, 54)**

Comparison (with Other Cases of Disasters) (4)

> Following the fire on the Piper Alpha oil platform in the North Sea in 1987 in which 167 men died, the official inquiry found numerous defects in the safety management system which had not been picked up in company auditing.
>
> **Hopkins (1999, 29)**

> ...and, one of the central conclusions of most disaster inquiries is that the auditing of safety management systems was defective.
>
> **Hopkins (2000, 54)**

Comparison (with Good Practices) (4′)

> BHP coal had clearly learnt the lesson of its previous auditing failure. It had understood that the hallmark of a good audit is that it must be thorough enough to uncover the bad news about safety and convey it upwards to the top of the corporation.
>
> **Hopkins (1999, 35)**

Comparison (with Safe Properties as Theorised in Research) (4″)

> The strategy which HROs adopt is collective mindfulness. The essence of this idea is that no system can guarantee safety once and for all. Rather, it is necessary for the organisation to cultivate a state of continuous mindfulness of the possibility of disaster.
>
> **Hopkins (2000, 140)**

> ...but in the case of Longford, safety auditing, an ideal opportunity to focus on the possibility of failure, was turned into an opportunity to celebrate success.
>
> **Hopkins (2000, 142)**

Recommendations ("Must") (5)

> A rigorous audit needs to examine the hazard identification strategy and make some effort to seek out hazards which may have been missed, so as to be able to make a judgement about how effectively hazard identification and control is being carried out.
>
> **Hopkins (2000, 86)**

Counterfactual Arguments (6)

> Had it reached the highest levels of the company and had been acted on, it would have averted the incident.
>
> **Hopkins (2000, 93)**

And this is all the more relevant that a single case can talk to practitioners far beyond its specificities. "People working in other companies in the same industry as the company about which I have written often tell me that many of my descriptions are applicable to their own company. I might as well have written for their own organisation, they say. This makes them exceedingly uncomfortable since they recognise that their organisation is at risk of suffering the same kind of accident" (Hopkins, 2016, 95).

As written above, many of the boxes in Figure 2.1, which gives a broad visual overview of the accident, could be opened and decomposed according to this narrative scheme. *Poor shift handover, failure of incident report system* or *operating in alarm mode* are commented in his book chapters using the same principles of describing, explaining, comparing, etc., which in turn provide the practical and persuasive strength of his writings.

But there is even more behind this visual map of the event because what shaped Hopkins mindset has not yet been discussed. It is now clear that his view is a system view, but this is not precise enough to enter into the intellectual sensitivity and rich interpretation of the author. When one does so, the acclaimed reception by practitioners of Hopkins' writings can come as a surprise considering that he started from the critical side of sociology.

Indeed, although both Perrow and Hopkins share their influence in Marxist sociology, one difference between the two sociologists is that Hopkins applies to safety a strong practical mindset and seems to be as happy, perhaps even more, outside academia than within it. He seems happier engaging with managers than trying to publish in top academic journals in sociology, which he describes as being biased by considerations that are of no interests outside academia (jargon, publishing strategies, etc.).

But first, to show the close connection to Perrow sociological sensitivity, and then how Hopkins completed substantially what Perrow started but did not pursue with the second thesis of *NA*, I need to proceed with a similar approach that in Chapter 1. After unravelling the ingredients of his successful storytelling, I need to reveal another face of Hopkins that is hidden so far to most readers, Hopkins the theorist, the unofficial theorist of the second thesis of *NA*.

Theory of the Second Thesis

The 1980s: Critical White-Collar Crime Model of Accident

So, if the principles of a narrative structure behind his success with practitioners is mostly unknown to many readers, the same applies to his sociological background. It is precisely because of this that Hopkins wrote a personal retrospective of his carrier (Hopkins, 2016), to express more clearly that it is the discipline of sociology which provided him with the methodological and theoretical tools to write about disasters the way he did.

Although very clearly expressed in his biography, I would like to add some insights which allow me to argue about his specific contribution as I see it. Hopkins began in the 1970s with a socio-legal and critical WCC orientation. In this reading of societies' injustice is intrinsically created by social structures that favour the powerful at the expense of the less powerful. Power can be material with wealth but can also be symbolic but is often the result of both, which allow elites to influence societies' institution for their benefits.

This injustice is translated in laws and a legal system made of tribunals, judges and trials which (re)produce the institutions needed to maintain the status of the powerful. Through conflicts, social movements and changes of institutions, societies can overcome this injustice. To study law in relation to how societies produce, enforce and sanctionit, it is a sociolegal angle that offers a particularly acute perspective on the dynamic of the social world.

To explore the topic of WCC is an opportunity to remain close to this inspiration of understanding some of the manifestations that reveal how societies work, and Hopkins investigated this area in the late 1970s: studies of WCC (or the crime of the powerful) target, the unlawful conducts (or misconducts) of people with high status, such as politicians or top managers of corporations.

A sociolegal view of this topic considers therefore the way the law (e.g. health and safety, consumer protection, environmental protection, accounting) is produced and designed, applied, enforced and sanctioned. A sociologist must, to do so, examine organisational and managerial practices to understand not only how the law is applied in practice, but also how it was produced in the first place, and then he/she must research how it is enforced, and beyond, how it is sanctioned in case of breaches.

This a program that requires the sociologist to collect data from a vast number of sources, from organisations to congress through agencies and tribunal practices. When Hopkins turned to safety in the 1980s with this mindset, he followed this program in a series of independent but connected articles that explored safety practices in companies and the origin of disasters, the constitution of health and safety law and its principles, the enforcement of the laws by agencies and inspectors and also judges' sanctions of breaches of the laws by companies (Hopkins, 1981; Hopkins, 1984; Hopkins & Parnell, 1984; Hopkins & Palser, 1987).

Let's comment briefly, simplifying a little the articles but without betraying their main messages. First, for Hopkins, behind accidents, one finds workers cutting corners under production pressures, translated in bonuses to reach production targets. His rational is that workers can't comply with the law if their working conditions do not allow them to or push them to do otherwise. This is a managerial problem and cannot be reduced to workers' behaviour, this is a second point (Hopkins, 1984; Hopkins & Palser, 1987).

Third, when the law is not properly complied with, but that inspectors who know about it do not enforce compliance, the system proves its limit and expresses its leniency towards owners of companies through co-optation of inspectors (Hopkins & Parnell, 1984). Fourth, when breach of law following a disaster is not punished by judges, the system again perpetuates the status quo that produces the conditions for accidents to repeat (Hopkins, 1981).

Taken together, Hopkins' articles published in the 1980s provide a framework to both conceptualise what can be called the critical WCC model of accidents. The critical view can be read (Box 2.2) and visualised as follows (Figure 2.6).

Accidents in the workplace are therefore very good examples of injustice for which owners of companies expose workers to risks because of their lack of concern for compliance with the law. And, if the law in the first place is not adequately designed to ensure safe working conditions, it might also be the sign of an attempt by powerful interests to limit its reach. In the case the law is adequate but not enforced or punished, it might this time indicate the co-optation of inspectorate or a class bias by the judges more favourable to owners than workers.

One now understands how compatible this view is with the second thesis of *NA* by Perrow, even if he is a sociologist of organisations, not a WCC researcher, because Perrow keeps a broad perspective and situates organisations in the context of the relationship between corporations and societies, best illustrated by his macrodepiction and comparison between error-avoiding (aviation) and error-prone (maritime) systems.

Derived partly from the Marxist critique of capitalism, both sociologists have a power view of the problem of accidents and safety. Workers are exposed to risks because interests create and maintain social structures that determine their working conditions. This, in fact, is a classic formulation for any critical sociologist who would study different topics, whether, let's say, business, safety, education or prisons.

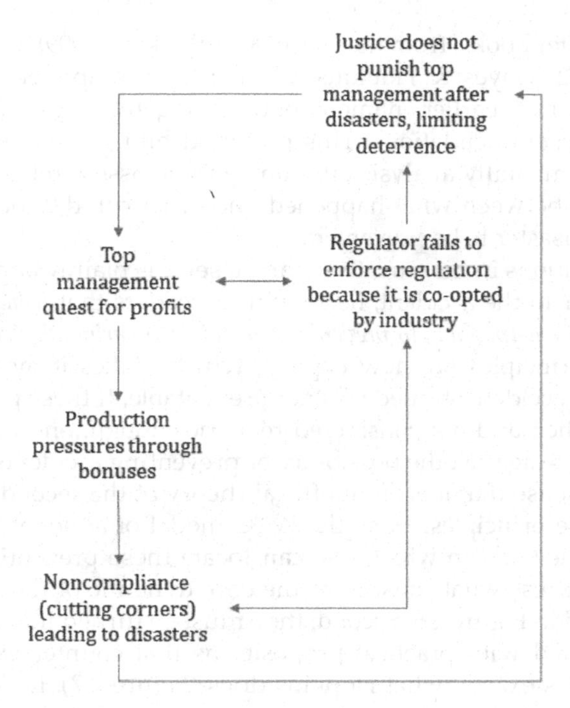

FIGURE 2.6
Hopkins' 1980s critical WCC accident model. WCC, white-collar crime.

2000s–2010s: A Normative Theory of Safety

But Hopkins does not stop there, he does not stop at the critical formulation. As explained earlier with the narrative structure, he complements this critical analysis with propositions of recommendations, moving from the critical to the practical for different issues identified, hence his success. Hopkins has also reflected upon the limits of the Marxist frame, "it was precisely because this was not a useful analysis that I had moved away from the Marxist framework" (Hopkins, 2016, 24), without abandoning his critical stance. This is why I noted earlier that the acclaimed reception by practitioners of his writings could come as a surprise considering that he started from the critical side of sociology.

Critical sociologists are not very often welcome or invited in for-profit companies by top managers and health and safety managers to advise them on their management practices, and many of his readers do not know about the critical view of capitalism which propelled Hopkins to study safety. He comments, "when talking to the most senior people in large corporations, I have often felt like an imposter, knowing how conservatively they think and therefore what a political gulf lies between us" (Hopkins, 2016, 7).

But books after books, from the late 1990s (Hopkins, 1999) to the mid-2010s (Hopkins, 2012; Hayes & Hopkins, 2014), Hopkins applied the narrative sequences defined earlier, namely putting together descriptions, explanations and recommendations. This provided him with a normative tool, evolving incrementally analysis after analysis, to assess retrospectively the degree of gap between what happened and what could/should have been done for the disaster to be prevented.

Because solutions in his narratives can be seen as plainly applicable by the company prior to the accident, he convinces readers that *it happened despite knowing that it was possible to prevent the accident in principle*. With Hopkins' work, these principles are now explicit, which makes it more difficult to argue that the accident studied was not preventable. If these principles were not made explicit and not considered realistic enough, one would be more entitled to argue against the possibility of preventing accidents.

It is in this sense that it is an unofficial theory of the second thesis of *NA*. What are these principles? First, the WCC model of accident (Box 2.2) provides the framework on which one can locate these preventive principles. They must address what constitutes the core dynamic of the critical model and visualised in Figure 2.6. Second, they must be turned into positive statements associated with practical propositions that counteract the negative patterns. This is exactly what Hopkins does (Figure 2.7). Let's briefly comment these principles.

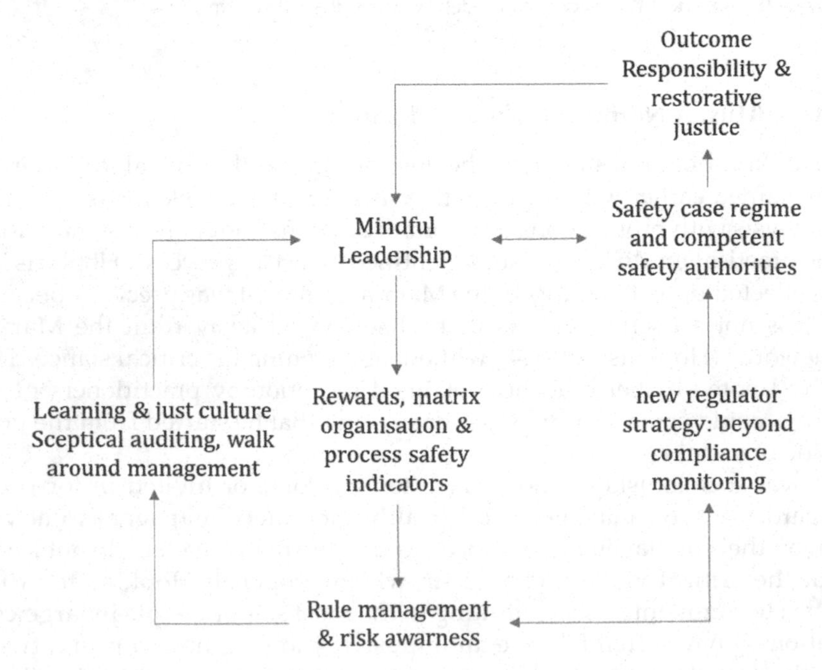

FIGURE 2.7
An unofficial theory for the second thesis of *NA*. *NA*, *Normal Accidents*.

Top management quest for profit (Figure 2.6) becomes mindful leadership (Figure 2.7), which must rely on walk-around management, learning and just culture but also sceptical audits (see Box 2.1 on problems with audits). Through careful attention to the details of operations, mindful leaders remain aware of concrete problems met in plants or sites. They do not trust the absence of problems to be a sign of success but an issue in the flow of communication, and they do not trust paperwork audit exercises.

Production pressure through bonuses (Figure 2.6) becomes reward scheme, matrix structure and process safety indicators (Figure 2.7), which are designed to ensure the balance between achieving production targets while not sacrificing safety. Structure of organisations should allow engineering and safety departments to be heard at the highest levels, top managers should not be financially rewarded (by markets) to meet short term objectives and indicators should refer to process safety, and not only occupational safety.

Noncompliance (cutting corners) leading to disasters (Figure 2.6) becomes risk awareness and rule management (Figure 2.7) because, first, operators need to know sufficiently the hazards they are dealing with or to know when to refer to engineers when unsure (risk awareness), and second, managers need to adapt rules when needed for adaptation of expert practices in the field.

Regulators failing to enforce regulations because of co-optation (Figure 2.6) become a new regulator strategy and safety case regulatory regime (Figure 2.7) in order to make sure that regulators require a company both to assess risk in order to design adequate prevention measures (safety case) and to verify if operational practices comply with the level of risk decided through these measures (inspection beyond compliance).

Finally, justice not punishing top managers and limiting deterrence (Figure 2.6) becomes outcome responsibility and restorative justice (Figure 2.7). Top leaders should be held accountable even when the complex structures of corporation make it possible for them to hide behind the argument that they didn't know (outcome responsibility), and they should be confronted with the suffering that they cause by meeting those who lost their loved ones in accidents (restorative justice).

This is a particularly elaborate and dense model of several features, which articulates many insights at different levels of social analysis. The sophistication of this model should not be underappreciated, however, as it requires users to be familiar with social sciences background. Of course, it is only when embedded in a specific case study with the narrative structure that this normative model of safety becomes highly appealing, persuasive and convincing to most readers.

Once articulated, they become an alternative, more positive and practically oriented version of the critical WCC model of the 1980s, replacing a description (Box 2.2) by another (Box 2.3) and a visualisation (Figure 2.6) by another (Figure 2.7). This is this model that I argue is a theory, a normative one, for the second thesis of *NA, accidents happen despite knowing that they could be prevented in principle.*

BOX 2.2 THE CRITICAL WHITE-COLLAR CRIME MODEL OF ACCIDENT (LINKED TO 1980S' ARTICLES)

Disasters happen because employees cut corners under companies' production pressures in the quest for profit incorporated in incentives to do so. One problem is that these safety violations which lead to such events are not punished beforehand by inspectorate following their visits because they are co-opted by the industry and do not enforce the expected level of rule compliance that the law and public would expect. Despite this, neither managers nor inspectors are prosecuted in the aftermath of a disaster for the pain they inflict to the families of the loved ones who perished at work, which also contribute to undermine the deterrent purpose of the law, and therefore the prevention of disasters. Moreover, behaviours of workers are often unfairly targeted by companies (sometimes regulators) as the primary cause of these events, whereas it is foremost a design and organisational issue which managers are entirely responsible for.

BOX 2.3 A NORMATIVE MODEL OF SAFETY

Mindful leadership must promote, with the help of carefully designed senior managers rewards schemes, matrix structures and process safety indicators, a risk awareness among companies' employees, also based on rule management. Internally, mindful leadership should rely on walk-around management and learning principles which include the handling of blame through just culture, and the need for adequate practice of audits, promoting sceptical principles. Externally, mindful leadership should be facilitated by inspections that go beyond compliance, principles of restorative justice and outcome responsibility.

Technology (Task), Structure, Goal and Environment

Although not detailing further the principles that Hopkins articulate in his model in this chapter, they frame his accounts of disasters and rely on the implicit assumption that it is what other companies, and sometimes industries, do to produce safely. Together, these articulated principles create currently one of the most integrative, normative and sophisticated safety models, bringing a sociological angle to the view, almost mechanistic, of safety management systems, as for instance, advocated in standards.

In Hopkins' rational, technology should be designed to prevent catastrophic events. Such design should be based on sound engineering standards and risk analysis (e.g. hazop), defining what decision-making processes operators must follow to remain within safe practices. Operators' tasks should be suitable to work with an understanding of choices made about the coupling between hazards, procedures, practices and human–machine interfaces (i.e. rule management and risk awareness, in Figure 2.7). What Hopkins addresses through rule management and risk awareness corresponds to Perrow's analytical categories of technology and task.

Structures of organisations have to be centralised to ensure that safety is granted enough weight in decision-making processes, at the highest level and throughout the entire organisation. This is a design choice that mindful leaders must apply and that should work in practice well enough. These structures should also be based on incentives that convey the right message about what is expected in terms of safety, compared to other potential conflicting goals of the corporations.

Structure and goal of organisations are in this respect tightly connected and not fully distinguished (more about this in Chapter 3). And in order to appreciate the level of achievement in practice of these design choices, learning should also be at the heart of how leaders manage their business, an issue that questions the quality of communication and flows of organisations within the corporations. Indeed, informal practices that can contribute to deviate from safe performance have to be identified through these learning processes, including reporting system and audits with the help of adequate centralised structures (i.e. mindful leadership, rewards, matrix organisation and process safety indicators, learning and just culture, sceptical audits and walk-around management in Figure 2.7). This is structure and goal in Perrow's model.

Organisations' environments have to be supportive by providing adequate standard of engineering and practices (industrial environment), regulatory regimes (legal and political environment), appropriate handling of top management by justice (judicial environment) and realistic return on investment expectations (financial environment) considering the hazardous processes involved. (i.e. beyond compliance monitoring, safety case regime and competent safety authorities, outcome responsibility and restorative justice in Figure 2.7). Of course, this environment is only partly in the hand of top management, but not entirely out of reach through lobbying and industry orientations, which are the products of their most influential members. This is Perrow's environment.

Hopkins' praise of high-reliability organisation (HRO) research tradition (this point is further developed later) is linked to this normative search for properties or principles that allow companies to defeat the technological determinism of Perrow, his tight coupling and interactive complexity (Hopkins, 2000, 2009). In Hopkins' mind, there are many successful companies around the world which show that it is possible to operate without major events. There is no technological determinism. Only if powerful actors of companies pay sufficient attention to safety, design appropriate organisations and dedicate sufficient resources, if need be and as often required, under the pressure of other powerful external actors (i.e. regulators), can accidents be avoided. There is no fatalism. This combined critical, organisational, managerial, legal and political view underlying Hopkins' approach is highly compatible with Perrow's own power lenses.

When doing so, it appears that classic themes of organisation theory as discussed by Perrow shape indeed Hopkins' own analytical choices. Topics such as structure, incentives, goal and leadership or environment of organisations are at the heart of Hopkins' narrative of disasters, and of his normative model. These topics correspond therefore to Perrow's decomposition of organisation into technology (and task), structure, goal and environment, as applied to high-risk systems, which he situates in broader context of societies, as described in Chapter 1. As a result, it is obvious that a common broad model of disaster and safety for both sociologists emerges and is represented in Figure 2.8.[1]

(Safety) Culture

It is interesting to mention that both Perrow and Hopkins have a rather distinctive relationship with the notion of (safety) culture and regarding its centrality in the sociological and safety discourse. As sociologists, they cannot deny the value and significance of the notion of culture to the analysis of social interactions, but they distance themselves from its centrality in two contexts. First, as explained in Chapter 1, Perrow wrote "we miss a great deal when we substitute culture for power" (Perrow, 1999, 380), to challenge Vaughan's emphasis on the interpretation of engineers over the ambiguities and uncertainties of components' behaviours in the Challenger case, at the expense of the pressure exerted by the NASA's managers on its subcontractors. Perrow's critical stance stresses power rather than culture, and his critics is directed at the sociologists of organisations who favour culture over

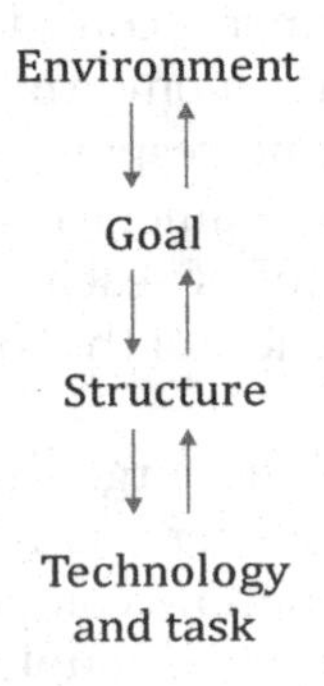

FIGURE 2.8
Perrow and Hopkins' common model of disaster/safety.

[1] Note that it is also compatible with Vaughan's distinction between environment, organisation, choice and cognition (Vaughan, 1999), despite, first, a stronger emphasis on goal in Perrow and Hopkins (more about this in the next chapter, when discussing strategy), and second, a proposition of refinement on situated cognition by Vaughan which is absent from Perrow and Hopkins.

power (see Chapter 1). Perrow does not target the concept of safety culture, but culture in general.

Second, Hopkins' critics of culture is more addressed to the way it is portrayed, used and defined in the industry and by many consultants in relation to the concept of safety culture (Hopkins, 2016). Safety culture has indeed become in the past 20 years a concept and then a business product on a safety market (Le Coze, 2019b), one that is often portrayed through front-line individuals' attitudes, practices and behaviours, and often independently of choices, orientations and decisions of top management.

Instead, Hopkins argues first that the words "safety" and "culture" should not be associated together in the expression "safety culture". Safety is one dimension of a culture, but a safety culture cannot be described independently or extracted from the broader organisational cultures (and subcultures) of companies. Second, Hopkins contends following Schein (Schein, 1992) that a positive culture when it comes to safety is a consequence of top managers' daily practices, style, problem resolutions, emphasis, incentives, etc. Culture is therefore a description, not an explanation, and what is needed is an explanation for why a culture is the way it is.

Hopkins considers that one determining factor in this respect is the degree of safety centralisation in the organisational structures of companies, which empowers safety expertise at the highest level (Hopkins, 2019). Culture conducive to safety comes from the top. Hopkins view of culture strongly derives from a power analysis of organisations, as Perrow does. So, although they value culture as a core analytical dimension of any social description, the emphasis of Perrow and Hopkins is on other equally important aspects of organisations which strongly influence and shape culture and safety, namely technology (and tasks), structure, goal, power and environment of organisations (Figure 2.9).

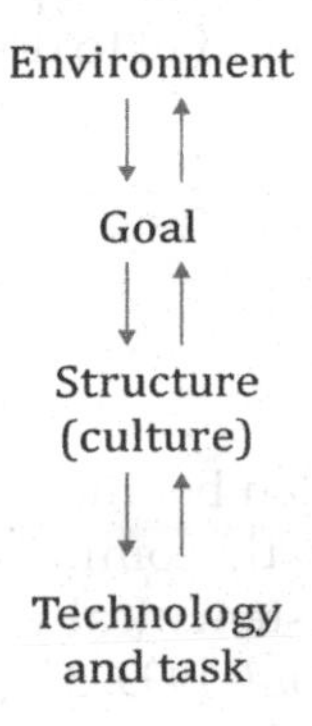

FIGURE 2.9
Culture as a product of environment/goal/structure.

Back to the Longford Case

Returning to the illustration at the beginning of the chapter, the Longford case (Hopkins, 2000) and visualising the causal diagram (Figure 2.1) drawn by Hopkins in 2000, one now understands that it is based on a substantial research development during the previous 20 years (1980s–1990s). This causal diagram is a critical view (although tempered by a practical side appealing to practitioners) that rejects the idea that the event was not preventable.

In this case, the regulatory and legal environment did not ensure sufficient scrutiny to criticise the absence of Hazop, whereas the market ideology of privatisation stopped government from initiating alternative supply of gas (this aggravated the consequence of the initial event). The financial environment of market forces led to cost cutting, which had an impact on the choice of organisational structure.

Indeed, engineering expertise was not available in the plant because moved back to headquarters, away from operations (e.g. an example of failing organisational structure to provide qualified support to front-line actors). Activities of operators on site suffered from such an absence, and informal practices developed without the possibility to correct their implications for safety (e.g. problem in rule management), before the accident caused by flaws in reporting incidents and auditing practices (e.g. lack of sceptical audits).

Confronting the Second Thesis of *Normal Accident*

With Chapters 1 and 2, it appears that together, the sociologists argue about what has been formulated as the second thesis of *NA*, *accidents are normal because they repeat despite knowing that they could be prevented in principle.* Although one could argue that it is quite a common sense and intuitive version of accidents which assumes that top management can favour profits over safety, namely that companies fail to do what people expect them to do (i.e. produce safely), this second thesis of *NA* has not been made explicit in safety research to the same extent that Perrow or Hopkins proposed. Why is this?

The Complexity Argument

First of all, this thesis was hidden behind the first and most famous thesis of Perrow, something he indirectly commented, "I would say that most of the work in the risk area is systematically detoxing the power aspects of my work" (Perrow in the *afterword*, 1999, 379). By referring exclusively to the engineering, technological and deterministic argument of *NA*, the maintenance programs, operational practices, learning processes, organisational structures, executive choices and regulatory controls (Figure 2.1) fell in the background although they are the best indications of how safe a system is.

But because of a mix of fascination, awe and modesty about our grasp of complex sociotechnical systems, Perrow comforted indeed in 1984 this inclination with his academic take on this argument. A recent version is found in Arbesman (2016), with an emphasis on software in the context of algorithms, machine learning and artificial intelligence. "The vast majority of computer programs will never be thoroughly comprehended by any human being (…) the werewolves of our time are the unexpected behaviour that lurch forth from the systems we build, the sinister embodiments of all the forces that have made our systems ever more complicated and less understandable" (Arbesman, 2016, 80, 93).

And it was extended and strengthened later in the 1990s by other authors as mentioned in Chapter 1, most famously Vaughan who argued in the case of the NASA that engineers and managers had to deal with uncertainties while under pressure to meet bureaucratic and political accountabilities. Following an interpretation first established by Rasmussen in cognitive engineering and system safety (Rasmussen, 1990, 1997) then empirically illustrated by Snook (2000), many safety writers have also popularised this idea to a wider audience of practitioners (e.g. Dekker, 2011; Hollnagel, 2014).

But this argument can be fallacious if one considers a good level of safety to be the product of choices by top decision-makers and of an appropriate degree of regulation exerted by states. This argument of complexity can indeed relieve leaders from their responsibilities and regulations by state from their centrality. If accidents happen from time to time despite our best efforts to prevent them, then leaders have a convenient explanation at hand to escape criticisms, but regulations too. This is, perhaps, another reason for the popularity of the original thesis (and its extended version) of *NA*.

A Sophisticated, Hidden, Normative Model of Safety

Second, Hopkins' important contribution to safety research has been so far considered mainly as an exercise of successful storytelling, hiding his theoretical input and hiding the fact that sociological works could also be oriented towards prescription. Venturing beyond the critical to elaborate the practical side of the sociology of safety, Hopkins developed over the years a (evolving) normative model projected into the past to compare what happened with what could or should have been done for catastrophic events to be prevented.

Third, there is a problem, in general, in social sciences with any normative statements not only because they take the risk of being easily challenged by reality but also because they imply a judgement about how things should be, which is against the ideal of value neutral analysis advocated by the detached scientists, but also professional sociologist. These are complicated topics and open a vast array of problems, and I discuss shortly three of them.

To start with, I think it is important to distinguish the critical view as developed by Perrow with the practical one by Hopkins. Both are normative, but the former remains quite judgmental and ideologically driven without really suggesting recommendations or implying them (e.g. stronger states, less profit driven leaders, decoupling of high-risk systems), while the latter is practical, explicitly suggesting ways of thinking about problems to solve them (hence his success with practitioners).

Next, I think that it is important in safety to be able to refer to normative models that help us think about data, either to sensitise data collection or to help think about what is realistic and what is not in terms of safety practices, but also why, in certain cases, despite not following what could be expected, companies do not suffer major events, in other words, that despite their imperfections, they still operate to sustainable level of safety in their social and political environment.

But, and finally, the last point raises an important question about the degree of realism of the normativity of the model. Because Hopkins (as much as Perrow) analyses events of the past, it differs from study of daily operations. The normative principles are indeed artificially articulated by Hopkins based on his knowledge of existing practices in industry, his own suggestions of recommendations or his use of research models (e.g. HRO). There is no certainty that his normative analysis is what companies, without experiencing major events, apply in daily operations. Hopkins' model is an articulation of various known practices or recommendations about how to improve safety.

The fact that Hopkins' narrative is persuasive does not mean that the suggested practices correspond to companies' practices. Perhaps, some of these practices are difficult to implement, perhaps that there are limits in rule management, sceptical auditing or matrix structures (Figure 2.7). To come back on the second meaning of normal (Table 1.3, Chapter 1), what real practices are in relation to a normative model remains a vexing problem. There are always observed imperfections when compared to established ideals, standards, norms or procedures. It is more appropriate for this reason to write of "daily" operations and then of "normal" operations.

For instance, Hopkins' argument on the role of centralised safety structures in company is one way of asserting an organisational version of how a culture conducive to safety is created (Hopkins, 2019). To make sure messages circulate from the bottom to the top of companies, safety should be centralised and available at senior management levels. Centralisation means that safety exists along other functions (e.g. maintenance, human resources, finance) with an equal weight or influence in the decision-making processes, at least formally.

This disposition is likely to maintain a balance in the organisation between safety and other dimensions. This proposition is an important one. In Hopkins' writings, it conveys an alternative discourse to the usual one associated with culture targeting individuals. It is a useful simplification targeting practitioners about the virtue of thinking safety concretely from a sociological perspective. One problem with this statement is that many safety-critical organisations can operate without such centralisation for a while without experiencing a major event (Le Coze, 2020).

To understand this gap between an ideal (e.g. centralisation) and reality, between a normative model and reality, an aspect such as operational constraints (e.g. cost cutting) needs to be introduced in our description and conceptualisation of high-risk systems. The reason is that a disaster is often the result of pushing an organisation beyond its limits. Centralised structure can help avert going beyond such limits. Centralisation is therefore a preventive design against the risk of losing a balance between safety and other goals. But pushing comes first from strategic decisions made by top management. So many organisations do not suffer major events even in the absence of centralised structures, because centralised structure is one aspect among others. This point is more thoroughly discussed next in Chapter 3.

When asking such questions about normativity and reality, one suggests that it might not be an easy task to situate perfection from imperfection, and we have now, at the end of Chapters 1 and 2, two options that remain opposed, advocated by different authors: on the one hand, the original (Perrow, 1984) and then its extended version of na (Vaughan, 1996; Snook, 2000), the first thesis of *NA* about the impossibility of preventing accidents in complex sociotechnological systems has strong supporters and is a popular version of accidents; on the other hand, the mostly hidden second thesis of *NA* that *accidents are normal because they repeat knowing in principle that they could be prevented*, because top managers' failure is supported by some but is not mainstream in the literature although a popular version of the "accident waiting" to happen (Perrow, 1999, 2015; Hopkins, 1999a, 2019).

A New Formulation of the *High-Reliability Organisation versus Normal Accident* Debate?

This of course is reminiscent of the influential *HRO versus na* debates of the 1980s and 1990s (e.g. Rochlin, 1993; Sagan, 1993), which culminated with heated discussions between La Porte, Perrow, Sagan and Rochlin (Perrow, 1994b; Sagan, 1994; La Porte, 1994; La Porte & Rochlin, 1994). In the 1980s, a group of interdisciplinary (organisational psychology, engineering, political science) researchers (Todd La Porte, Karlene Roberts, Gene Rochlin) studied ethnographically what they then defined as "high-reliability organisations", namely organisations operating in unforgiving complex environments while performing "nearly error-free operations" (i.e. navy aircraft carrier, air traffic control, nuclear power plant).

They published their results in several articles (Rochlin, La Porte, & Roberts, 1987; Roberts, 1989; La Porte & Consolini, 1991), and this became an influential research tradition over the years (see Le Coze, 2019d for an overview). HRO researchers described a number of positive features about tasks, structure, training, culture, learning, decision-making and leadership that they observed in the daily practices, which helped explain the conundrum of a high level of performance despite what they considered to be operations under very trying conditions. Let's define shortly these positive features.

Tasks were redundant within team to capture and recover from potential unnoticed mistakes; structure was underspecified to allow fast problem resolution to migrate towards expertise rather than following hierarchy when required; training relied on strong socialising processes emphasising safety as a paramount goal; culture infused people with a sense of unease about operations that made them alert of small discrepancies before they could escalate; learning favoured the expression of mistakes without the fear of blame; decision-making relied on a broad perspective of operations by a leadership that created these favourable conditions for culture, training, learning and decision-making to respond to the very trying conditions described by the HRO researchers.

In the 1980s and early 1990s, na and HRO were two parallel research programs framed from different perspectives at similar times, both following Three Miles Island in 1979 (Perrow, 1982; La Porte, 1982). But this changed when, Sagan, a political scientist looking for a theoretical background for his study on nuclear weapons, opposed explicitly what he described as two schools. For Sagan, on the one hand, HRO researchers believed in the possibility of preventing disasters and looked for these properties through fieldwork.

On the other hand, na argued the opposite. This opposition was not well received by the HRO researchers; both La Porte and Rochlin did not recognise themselves in Sagan's framing of their insights in relation to na (La Porte, 1994; La Porte & Rochlin, 1994). "I do not think the 'optimistic/pessimistic' dimension is a particularly useful one (…) our efforts are complementary to Perrow and other contributors to the 'normal accident' perspective, certainly not a competing theory" (La Porte, 1994, 209, 210).

They did not see themselves looking for recipes to design and manage high reliability organisations as Perrow's commented. "There is something faintly 'B-school' about the High Reliability Theory list; no one can be against clear safety goals, learning, experience and so on" (Perrow, 1994b, 215).[2] They rejected this interpretation. "He obfuscates our work with the invidious term 'B-school'" (La Porte & Rochlin, 1994, 221) and added "it is folly to suppose, as many engineers and some managers do, that one can expect to design systems and operate them in failure-free modes for extended period" (La Porte, 1994, 209).

They emphasised instead the extraordinary demands such operations created for operators and targeted the lack of appropriate lenses, including vocabulary and concepts in organisation theory, to express these realities adequately. In short, these were "high reliability seeking organisations"

[2] 'B-school' for Business school.

(Rochlin, 1993), and HRO researchers mostly agreed with na, and did not see any contradictions with their studies.

But Perrow and Sagan (Perrow, 1994b; Sagan, 1994) maintained their positions about the HRO research despite the rebuttals by La Porte and Rochlin (La Porte, 1994; La Porte & Rochlin, 1994). First, Perrow and Sagan considered HRO to be missing a power view of organisations, something that Perrow precisely believed to be Sagan's most valuable contribution to his own version of na. "Group interests and power pervaded my 1984 book, but I did not note that many organizational theories were inattentive to not only bounded rationality, but to interests and power" (Perrow, 1994b, 217).

Second, Perrow regretted the lack of greater empirical acknowledgement for the role of environment of high-risk systems in the creation and maintenance (or not) of safe performances. He wrote about the HRO research limitations that "the main effect of the opening environment in HRT is that it demands safety, but otherwise no great effort is made to theorize environmental effects upon the operating system" (Perrow, 1994b). He then suggested a possibility of connection between his own macroperspective with those of the HRO through his characterisation of error-prone versus error-avoiding systems (Table 1.1, Chapter 1).

These two criticisms, the lack of enough consideration for power and environment, were translated in the following quote: "few managers are punished for not putting safety first even after an accident, but will quickly be punished for not putting profits, market share or prestige first" (Perrow, 1994b, 217). First, power of top managers and, second, environment aspects such as failure of justice and influence of markets were translating, in this sentence, his remarks about the limits of the HRO studies. These two criticisms make now of course a lot of sense in light of the second thesis of *NA* introduced in Chapter 1.

But there was a third comment. Perrow did not seem to believe HRO to be extraordinary organisations, and he thought that they exhibited similar properties found in other organisations. "When I think of the rolling mill operation of steel plants I studied in the 1960s, they seemed every bit as dangerous, as decentralized and as imbued with a safety culture as the prime exemplar of high reliability organisations; the aircraft carrier deck with its intense activity (...) In this sense, the sobriquet 'high reliability organisations' probably applies to more organizations than HRT would expect. Some industrial units are run at the intense rates" (Perrow, 1994b, 215). For Perrow, the HRO research program was flawed in this respect.

In other words, considering now together the three criticisms, organisations are safe, reliable and performant not because of unusual properties that organisation theory fails to explain, but when top managers dedicate (helped or pushed by their environments) enough attention to the issue of safety. This is a premise that is clearly shared by Hopkins as explained in this chapter. Hopkins has in this respect an instrumental use of HRO, seeing in their descriptions an argument about the possibility of preventing events if enough attention is granted to the problem of safety by powerful people. He incorporates thus many of the insights of the HRO studies when articulating topics such as learning, culture, structure or leadership (Figure 2.7).

These three criticisms should be seen as complementary rather than a rejection of the HRO project, although this complementarity slightly reorients indeed descriptions of high-risk systems. First, Perrow and Hopkins bring much more clearly the presence in the safety field of a critical and power view of organisations, something that is almost never really associated with the original debate of HRO versus na because of its focus on the first thesis of *NA* and its technological rationale of tight coupling and interactive complexity.

Here instead, accidents reveal powerful interests of leaders of companies or multinationals which societies can fail to curb through adequate environments (including regulations), not complexity of sociotechnical systems. As Hopkins made it clear in his rejection of the first thesis of *NA*, "despite the technological complexities of the Longford site, the accident was not inevitable (…) As the commission said, measures to prevent the accident were plainly practicable" (Hopkins, 2000, 5). This view implies some links to wider, political economy considerations (to be discussed in Chapter 4) and opens a different route to our understanding of safety, as a product of the relationship between business and society (Birch et al., 2016).

Second, if Hopkins' work includes the model of *collective mindfulness* built in the late 1990s within the HRO tradition to strengthen its theoretical basis (Weick, Obstfled, & Sutcliffe, 1999), he also incorporates many other sources of reflection (Hopkins, 2005, 2007a, 2012, 2019). This led him to develop an increasingly sophisticated, integrative and broad model of safety over the years (see Figure 2.7), which articulates findings from a variety of fields, organisation, regulation, management, cognition and engineering that translates his extension of the HRO lenses.

Third, and deriving from this second point, he targets explicitly (along with Perrow) top-level decision-making processes, approached through the notion of *mindful leadership* (Figure 2.7), which was not at the heart of HRO (Sutcliffe, 2018). He adapts therefore the concept of *collective mindfulness* through the notion of *mindful leadership* (Hopkins, 2007a). This focus on higher levels of decision-making is welcome because, as it will be argued next with the topic of strategy (Chapter 3), more needs to be done to better include top managers' actions and decisions, and their influence on safety.

Fourth, Hopkins' work reflects some important trends associated with globalisation in the landscape of high-risk systems (a point to be developed next in Chapter 4), which are not addressed in the *HRO versus na* debate. Because of the lack of emphasis on the environment of organisations (see above) and because globalisation was not as prominent in the 1980s, it was simply not part of the scope of the first HRO studies (Schulman & Roe, 2018). Perrow wrote subsequently beyond the HRO versus na framing after the 1980s and 1990s with his book dedicated to the next catastrophes (Perrow, 2011b).

Although it has not been introduced so far in Chapters 1 and 2, this book explores the interactions between natural, technological, cyber and terrorist threats and constitutes a strong indication of a change of risk profile in the landscape of high-risk systems expanding scope, scale and timeframe (to be introduced in Chapter 5).

Summary of Chapter 2

Hopkins' work supports the second thesis of *NA* by Perrow; he also considers most disasters to be preventable. Moreover, he develops further what Perrow did not. First, he writes in a way that persuades readers of the failure of companies. His persuasion rests on a narrative structure that combines multiple sequences of description, explanation and recommendation repeated across chapters of his books. Second, the themes of his chapters are articulated with a sociological interpretation of capitalist societies through the lenses of a critical view of the powerful. In this view, interests of a dominating class are maintained through a legal system that only evolves in the direction of less injustice through social movements.

This idea, explored through an organisational analysis of WCC and corporate crime, transferred in the field of safety, connects Hopkins with Perrow, the Perrow of the second thesis and power view, but moving from the critical (Perrow) to the more practical (Hopkins). Third, over the years, case after case, Hopkins builds a refined and integrative model of safety based on several analytical angles and layers of organisational analysis. He conceptualises safety through a combination of appropriate risk awareness, rule management, indicators, incentives, structure, culture, learning, auditing, leadership and environment (industrial, legal, financial and judicial). This is also compatible with Perrow's own categories (introduced in Chapter 1) of technology and tasks, structure, goal and environment of organisations.

For this reason, Hopkins is the unofficial theorist of the second thesis of *NA, accidents are normal because they repeat despite knowing that they could be prevented in principle.* Accidents are preventable. Finally, his formulation pursues the *HRO versus na* debate of the 1990s by bringing a combination of critical and practical views together, by integrating many different sources into a broad normative sociological model, by targeting top decision-makers (and regulations), and by tackling issues which reflects some trends affecting the landscape of high-risk systems. All these points constitute the bridge to the next chapters of *Post Normal Accident.*

3

Errors from the Top

Introduction

With Perrow and Hopkins, one can rely on powerful lenses to look into disasters and safety of high-risk systems. Whereas Perrow is critical and does not provide in-depth accounts of single cases, Hopkins goes deeper and builds a normative model to persuade readers about the possibility of preventing major events. The argument of this chapter is that one needs to further sensitise and appreciate the complexity of operating high-risk systems, and to do so requires to pay attention to an underappreciated issue in safety, strategy.

As explained in Chapters 1 and 2, one important contribution of *Normal Accidents* (*NA*) and now, *Post Normal Accident* (*Post NA*), is indeed to move away from human error analysis of accidents by considering several features of high-risk systems (technology tasks, structure, goal and environment) while therefore shifting the focus of the analysis from the bottom to the top of organisations, and their environments. A core message of both *NA* and na is that a human error of a process plant operator or of a pilot is not an explanation when large-scale events occur.

A human error of front-line actors leading to a catastrophe is most of the time linked to deeper organisational and regulatory issues in turn connected to choices made by top decision-makers, namely companies' strategies. One could say that an interest in strategy derives quite evidently from the analysis of key sociological works in the field of safety. Both Perrow and Hopkins, who have been extensively reviewed in the Chapters 1 and 2, could be said to address to a certain extent the topic of strategy.

Yet, the argument is that it should be more explicit. This chapter contributes to a more explicit inclusion of strategy in the safety discourse and in our account of disasters. *Post NA* endorses the idea that disasters, and conversely, safety, cannot be studied without a dedicated attention to strategy. Whereas human error of front-line actors cannot be the end of the story, the same cannot be said about strategic errors by top leaders. They often are the ones that are the most important to understand, then, if possible, to prevent.

This chapter starts with showing that strategy has been approached by Perrow and Hopkins, but with a rather critical angle, other options are possible and suggested by other writers, Starbuck and Farjoun in particular. Second, this chapter connects the literature on strategy with safety research to offer a way of combining the two. Third, three case studies are introduced to illustrate the *Post NA* proposition that strategy mistake, failure and fiasco should be at the heart of our understanding of the operation of high-risk systems.

The Obviousness of Strategy

Failing Executives and Corporate Malfeasance

The idea that strategy matters for safety is at the heart of the second thesis of Perrow. By combining dimensions such as technology, structure, goal and environment that derived from his sociological approach of organisations, Perrow considers strategy to be key to safety performance, although he more often formulates the notion with goal in his vocabulary. "Structure, power, goals, environment, these are the concepts that have been stressed in this book" (Perrow, 1970b, 175). In fact, strategy, at the time Perrow developed his view of organisations in the 1960s and 1970s, was not yet an established research field (Carter, Clegg, & Kornberger, 2013) but was already central to his analytical mindset: "for both the social scientist and his management trainee, the most complete understanding of an organization will come through an analysis of its goal and basic strategies" (Perrow, 1970b, 180).

Based on the analysis provided in Chapter 1, it has become clear that Perrow situates the production of safety at the top of organisations and how networks of actors and institutions constrain goals of companies to remain or not within boundaries of safe practices (i.e. error-prone versus error-avoiding systems). Commenting about British Petroleum (BP), a company that experienced several disasters within a short period of time (2005–2010, Bergin, 2012), he writes "a few years ago, an aggressive chief executive officer (CEO) at the British company ordered drastic cuts in operating expenses" (Perrow, 2015, 210).

Indeed, Perrow often refers to "executives not trying very hard" in a mild version, but sometimes also to "executive malfeasance" in a more radical version devised to "emphasise that more than a failure of proper executive behaviour is involved" (Perrow, 2011b, viii). This interpretation is quite consistent with the underlying principle embraced by Perrow that safety is almost antithetical to capitalistic systems that are built for profit.

There is therefore in his eyes an imperative for the presence of strong states, of laws favouring alternative sources of powers inside companies (e.g. unions) and of adequate regulations in order to counterbalance this search for profit because "almost every major industrial accident in recent times has involved either regulatory failure or the deregulation demanded by business and industry" and "the poster child is British Petroleum, though it is far from the only one" (Perrow, 2015, 210).

As explained in Chapter 2, Hopkins is close to Perrow, although more nuanced or at least not as explicitly judgemental because of his practical ambition. He shares with Perrow a background in critical sociology (Hopkins, 1981, 1984), but one that relies on detailed analysis of disasters, more than Perrow does. For instance, he does not describe Browne, the CEO of BP, behind the company's strategy as "aggressive".

He writes instead "cost cutting at Texas City was driven by a relatively low return on investment. Shareholders require capital to be invested in such a way as to maximise the return on that investment. This is the logic of the capital market. Companies that ignore this logic will fail" (Hopkins, 2008, 82). Strategy is described in this as dictated by markets (see the following).

Let's recall briefly his contribution detailed in Chapter 2 from the angle of strategy. Analysing events retrospectively on the basis of investigation reports and hearings, Hopkins has been referring prominently over the years to, among others, issues such as lack of mindful leadership (Hopkins, 2007a), inadequate incentives favouring profits over safety and the absence of relevant process safety indicators (Maslen & Hopkins, 2014) and, more recently, organisational structures in particular in relation to centralisation of safety (Hopkins, 2019) as underlying causes of disasters.

These studies are directly connected to strategic levels of companies, but not formulated as such in Hopkins' writings. For instance, coming back on the graphical representation of his analysis of Longford used in Chapter 2 (Figure 3.1), strategy is not mentioned in any of the causal boxes. Market forces at the societal level exert influences on cost-cutting and control failure of the company, but the notion of strategy is not formulated. It is somehow implied.

Indeed, Hopkins' emphasis over the years on features such as incentives, design of organisational structures, indicators of high-risk systems and mindful leadership is an explicit consideration of some aspects of strategy because these features are shaped by top managers who have indeed as their core attributes to make choices of organisational design, incentives

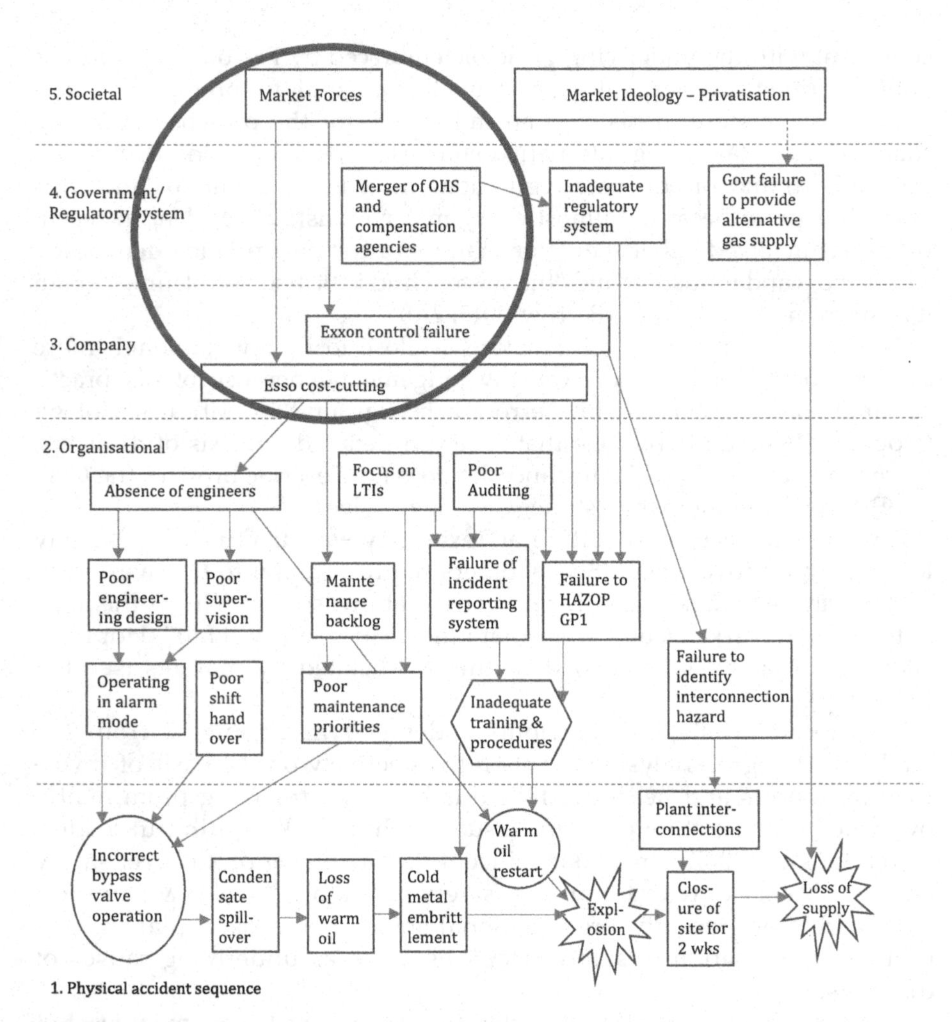

FIGURE 3.1
Elements of strategy in Hopkins' accident diagram (Hopkins, 2000).

(monetary, symbolic) but also to influence management style within the company.

Refraining from formulating moral judgement (see the quote mentioned earlier), he, moreover, acknowledges the multifaceted aspect of complex organisations, including their environment and, particularly, regulations. His normative option is primarily to design adequate organisational principles for safety and regulatory regimes that can curb the intrinsic search for profit through mechanisms favouring compliance, playing on the pluralistic range of economic, moral, legal and reputational motives, which should compel senior executives of companies to do as much as possible to remain within the boundaries of safe practices (Hopkins, 2007b).[1]

Organisations at and beyond the Limits

Whereas for Perrow or Hopkins (with the nuances indicated), the interpretation of accidents and safety derives from a critical view that capitalist search for profit is to be linked somehow and ultimately to executives' sacrifice of safety over production, the interpretation of Starbuck, an organisation, decision and strategy scholar, derives more from a psychocognitive interest in the activity of top managers (Starbuck & Milliken, 1988b; Starbuck, Greve, & Hedberg, 1978; for an overview, see Starbuck & Farjoun, 2007). The starting point is different than of Perrow or Hopkins because "good and bad results may arise from similar processes (…) the processes which produce crises are substantially identical to the processes which produces successes" (Starbuck & Milliken, 1988b, 39, 40).

Following this principle, Starbuck warns in the 1980s against the hindsight bias "retrospection wrongly implies that errors should have been anticipated" (Starbuck & Milliken, 1988b, 40), a position still advocated more than 20 years later. "Analysts have to surmount their tendencies to know more than they could have known, and they must formulate prescriptions that help decision makers to operate effectively amid complexity, uncertainty and ambiguity" (Hodgkinson & Starbuck, 2012). Starbuck conceptualises executives filtering the world in relation to complexity, ambiguity and uncertainty, quoting as an illustration the chairman of IBM who declared, in the period following World War II, "I think there is a world market for about five computers" (Starbuck & Milliken, 1988b, 39).

The future is not determined, and executives, top managers and leaders should be granted a more cautious retrospective appraisal in the light of failures (depending, of course, on their extent and nature, more about this in the following). Introducing with Meyer the interplay between ideology, business strategy and power, Starbuck describes also the sociopolitical nature of top decision-making beyond its cognitive dimension (Meyer & Starbuck, 1993).

Once translated with Farjoun into the concept of "organisations at the limits" (Farjoun & Starbuck, 2007), executive failures are understood in the light of a core tension in companies between "exploration of new possibilities and exploitation or protection of current assets". They add that "one comes

[1] Although more referenced for the perspective of a normalisation of deviance by NASA engineers about behaviour of technical artefacts (Vaughan, 1996, 2005), Vaughan also introduced the associated idea of a "trickle down effect" to conceptualise how choices made by powerful actors within and outside the agency (politicians in Congress, top administrators) translated in problematic safety contexts and constraints (Vaughan, 1997). She writes, "top administrators must take responsibility for mistake, failure, and safety by remaining alert to how their decisions in response to environmental contingencies affect people at the bottom of the hierarchy who do risky work" (Vaughan, 1997, 96).

usually at the expense of the other" (Farjoun & Starbuck, 2007, 558), which entails a degree of risk taking as an intrinsic aspect of strategy. Farjoun and Starbuck reassert the complexity of the issue because "People and organizations do not always know how far they are from the true limits or the extent to which limits are elastic, relative, or arbitrary. Therefore, progress in general, and exceeding limits in particular entails ambiguity, risk and uncertainty" (Farjoun & Starbuck, 2007, 543).

But they also leave no doubt to the reader about the roots of problems when dealing with technology and their environment "of course, since organizations choose their environments and technologies, the issues is partly one of business strategy" (Farjoun & Starbuck, 2007, 553). In this respect, they formulate the following interrogation: "If an organization is striving to exceed some of its limits, some people are pushing for very extreme performances. Who is exerting such pressure, and for what reasons?" (Farjoun & Starbuck, 2007, 562). Although not explicitly stated as such in their article, one can infer that they pursue two lines of interpretations. One is more critical: "organizations may be driven to try to exceed their capabilities by senior executives who pursue unrealistic goals because of insecurity, ambition, greed, hubris, jealousy or competitive zeal" (Farjoun & Starbuck, 2007, 554).

The other is a milder form of interpretation: "organizations may promise too much and overstep their capabilities, not so much as a result of conscious design but as response to cumulative flows of events or as unintended by-products of decisions and actions" (Farjoun & Starbuck, 2007, 545).[2] These two options will be used in a later section to interpret three case studies of safety as strategy (Table 3.6). From a safety point of view and the operation of high-risk systems, Starbuck has mostly published on Challenger and Columbia (Starbuck & Milliken, 1988b; Farjoun, 2005; Starbuck & Farjoun, 2005, 2007).

This presentation offers an overview of the current state of understanding of how strategy contributes to safety and disasters. Authors start from different assumptions, one more critical (Perrow), another critical and practical (Hopkins) and another more descriptive and managerial (Starbuck), but all agree on the centrality of this aspect. But they are cautious. Safety and disasters are multidimensional phenomena; they result from a complex interaction between technology (task), structure, culture, goal, environment, so to single out strategy from this complexity is difficult. Perrow is all too aware of this issue.

He contends that "classifying something as an executive failure rather than a mistaken executive strategy or a poorly performing executive, or even a failure by management or workers, is controversial. Observers can disagree,

[2] This reminds Vaughan's similar formulation introduced in Chapter 1: "no fundamental decision was made at NASA to do evil, rather, a series of rather seemingless decisions were made that incrementally moved the space agency toward a catastrophic outcome" (Vaughan, 1996, 410).

and since it is easy to have a mistaken strategy or a poorly performing organization, it takes a great deal of evidence to make the case for executive failure" (Perrow, 2011a, 293).

But what is interesting with these authors is their emphasis on other kind of actors who are traditionally studied in safety studies (e.g. operators, pilots, nurses, engineers). They characterise instead other influential actors at the top of companies, which can be connected to strategic literature. Safety and strategic research have not been much linked, excepted for the National Aeronautics and Space Administration (NASA) with attempts following Columbia (Farjoun, 2005).

Farjoun writes, to precisely indicate his focus on strategy, "I am concerned with why NASA and its leaders did not respond to accumulating global signs to imminent organizational failure as much as with why they failed to deal with specific operational issues" (Farjoun, 2005, 61). How is it that top managers are likely to make mistakes or errors when it comes to designing, choosing and selecting options that considerably shape operational contexts of high-risk systems?

What We Know about Human Errors

The problem of human error has been mostly studied from the perspective of front-line operators (pilots, surgeon, operators, nurses, soldiers, etc.) by disciplines such as cognitive psychology but also cognitive engineering and the source of a very rich literature in the past four decades (Reason & Mycielska, 1982; Reason, 1990; Rasmussen, 1990; Woods et al., 1994; Hollnagel, Woods, & Leveson, 2006; see Le Coze, 2015, 2019d for an overview). Three main insights can be retained from this literature. First, one should decouple errors from their consequences when studying disasters. Errors by front-line operators are not sufficient explanations because far more is involved in the breakdown of sociotechnical systems. Understanding of errors very often enmeshes description and moral judgements together because it is highly problematic to disentangle these two aspects (and the legal one too).

It is for this reason that decoupling errors from their consequences helps loosen the difficulty of unravelling the moral, operational and factual dimensions. And this proposition is closely linked to a second insights of research on errors at the sharp end. Studies of cognition have indeed revealed that error and performance are the two sides of the same coin. Errors are expected products of intrinsically fallible minds. But the opposite is equally true. The opposite view insists instead on the positive side of cognition, on the ability to cope with complexity, on the expertise backed up by years of practice, and on the ability to very often recover from errors and to adapt to unexpected situations. Moreover, errors appear to be intrinsic to learning processes.

One methodological implication is that work of front-line operators should be observed in daily operations, and not only following events that show how errors have had detrimental consequences. Applying retrospectively normative expectations is always to be done carefully. Thus, abstracting errors from their complex situated practices does not, very often, give justice to the level of expertise needed to perform their task. Only empirical studies of their daily performance can show it.

Note that this focus on both the negative and the positive sides of cognition must be understood in material, informational and team (or collective) processes. This is a third core insight of the study of error in safety research. Strengths and weaknesses of cognition are embedded properties that cannot be studied in isolation from their contexts. Isolated individuals at work without any colleagues and no mediatised environment (e.g. computers, tools) are never met in high-risk systems.

These three important insights apply well to front-line operators. Decoupling errors from their consequences (in retrospect), emphasising the positive side of operators during daily practices (and not only the negative) and recognising their embedded properties in contexts have strong advantages when it comes to learning and promoting safe performance.

A similar posture is surely granted for top decision-makers, as proposed by Starbuck earlier, although, of course, one cannot apply entirely the same principles. One difference which is argued in this chapter is, when it comes to disasters, that the link between errors and their consequences is causal in a different way in the case of strategy. It is indeed the choices made by top decision-makers which create the conditions in which catastrophic events occur. Front-line actors only trigger the chain of events that led to the outcomes.

It is the engineering, human and organisational conditions created by top management, by their strategies, which makes it possible for front-line errors to end up in disasters. So ultimately, disasters are the products of strategic orientations, which turned badly. This strong claim, which is also an important analytical complement to existing studies in safety research including Perrow and Hopkins, is furthermore advocated with the help of the strategic literature, to which I now turn.

Safety as Strategy

The Importance of Strategy for Businesses

We are quite familiar with the topic of strategy through the general media. We know that a product's successful development (as well as its decline) has much to do with strategy. We are also quite familiar with the topic of strategy in the context of companies whose products once dominated their markets but have since disappeared because they were unable to match the technological innovations of other companies.

Apple and Samsung's smartphones are examples of successful strategies, whereas the cell phones of Nokia, Ericsson and Blackberry could be considered strategic failures. The basic but central idea that strategies can shape the fate of organisations and their employees, as the losers or winners of today's global market competitions, is quite familiar.

The successes and failures of companies are indeed believed to be in the hands of their CEO, and the names Steve Jobs, Elon Musk and Bill Gates are revered worldwide (Guthey, Clarke, & Jackson, 2009). Because of their positions of power at the top of organisations, they ultimately determine a great number of orientations that indeed directly constrain the conditions of operations, and therefore their results. With their power comes the possibility of taking decisions and shaping organisational features with far-reaching consequences for the fate of their companies, their employees and also their customers (Kaiser, Hogan, & Craig, 2008).

Linear, Adaptive and Interpretive Views of Strategy

But how has strategy been studied? It has been the topic of research for three to four decades now (Mintzberg, Lampel, & Ahlstrand, 2009), and one can refer to an enduring distinction of the 1980s which differentiates linear, adaptive and interpretive approaches to strategy (e.g., Mintzberg, 1973; Chaffee, 1985; Johnson, 1987). The first one is grounded in the idea that executives can objectively describe the opportunities, resources and constraints of their business (inside and outside the firm). Strategy derives from this analysis, which is then implemented following instructions from the top. This is the linear view of strategy, a sort of translation at the executive level of the Taylorist ideal.

The second one is the adaptive version, stressing instead the experimental and incremental nature of strategy: companies evolve on a regular basis by introducing variations that keep them in line with markets, consumers and technological changes. "Employees might know much more about customer needs, business operations and inimitable practices, how to improve them, than a detached management" (Carter, Clegg, & Kornberger, 2013).

These variations are produced by the companies through small moves that can be initiated, tested and then adopted in several corners of the organisation, without necessarily involving top management at first. This is also described as the emergent view of strategy. The third is the interpretive perspective. It relies on careful attention to the cognitive processes, mental representations or constructs that top people in organisations use to make sense of their environment, companies' capabilities and causal relations between strategy and (opportunity of) success.

TABLE 3.1
Linear, Adaptive and Interpretive Views of Strategy

Interpretation	Short Description
Linear	Executives establish the strengths and weaknesses of the company in order to elaborate a strategy and then design the path to reach the desired outcomes, implemented by employees who follow this plan.
Adaptive	Strategy is the product of experimental and incremental changes (not necessarily planned) adapting the organisation to its environment; as a result, companies manage to remain successful businesses. These moves can be initiated at different places, not exclusively by top management.
Interpretive	Strategy is the product of complex cognitive processes trying to make sense of the world, creating power games about which interpretations within the cultural background of business leaders should prevail, considering the uncertainties of the market, consumers and technology developments.

It argues that strategy is often a question of competing visions within a firm which are shaped by power struggles as well as cognitive and cultural framing. Bold rather than incremental moves, or even disruptions in company strategies, might result, sometimes too late for the survival of a company and reflecting tensions, power struggles and coalitions behind strategic options. Table 3.1 summarises this classification.

Is strategy a product of top managers who decide through rational means of analysis about the strength and weaknesses of their organisation in relation to product developments and market potential? Do they decide then transfer the implementation to employees? Does structure follow strategy? The answer is yes for the authors belonging to the linear (and normative) view above, less so for the two others. In any case, decisions of orientation remain in the hands of such decision-makers, which leads to the power issue.

Power of Executives and Top Managers

Executives and top managers interact with a variety of actors inside and outside the organisation, from top management teams to the board of directors, within complex governance structures and networks (Gomez, 2018). Research in strategy has greatly contributed to untangling these complexities (e.g. Eisenhardt, Furr & Bingham, 2010; Johnson, Whittington & Scholes, 2013; Vaara & Whittington, 2012; Rouleau, 2013). It has put into question the taken-for-granted idea in the field of strategy that executives possess power.

Contrary to this idea, executives and top managers are in many ways constrained in what they can and cannot do. Boards of directors, regulators, shareholders, auditors, nongovernmental organisations and unions are examples of key negotiating partners which limit and constrain them in the exercise of power (to which we could add the materiality of technology or ecosystems). Moreover, power is not an individual attribute but a relational one, structured by specific social contexts.

Circumstances can bring more or less power to individuals, and an important part of an executives' job is in fact dedicated to manoeuvring to get as much power as possible to implement their ideas. Carter, Clegg and Kornberger consider, for instance, Machiavelli's view of power to be the most appropriate lens to conceptualise the field of strategy (Carter et al, 2013).

But other approaches see strategic success from an evolutionary perspective which also challenges executives' power or influence. According to this view, executives are only successful when selection processes favour their business. A successful strategy is not a result of how lucid and prescient executives and top managers have been but of external forces which they do not really control. No one can foresee the future and the complex relationships between customers' evolving preferences and competitors' moves. This evolutionary perspective undermines notions crediting top management and their power because luck is involved through this unpredictable selection process.

Another view is that executives' actions are only one aspect of why organisations succeed. As described earlier (adaptive view), ideas that turn into successes can derive from initiatives from different corners of the organisation and not exclusively from executives and top managers. Strategy is more distributed. Moreover, implementation of strategic orientations requires a level of implication, adaptation and improvisation by employees that powerful company actors do not necessarily anticipate but are key to meet expectations.

So, to attribute success solely to executives is a reductionist account of a more collective achievement: "the strategy we assume we observe is a result of an assemblage of practices. It is only the process of objectifying and reifying these practices that leads to what we think as objects" (Carter, Clegg, & Kornberger, 2013). So, power should not be overestimated, but the reverse is also very clear: executive power should not be underestimated. As Perrow contends, "organizations are imperfect tools, but the powerful ones appear to do well enough for their masters" (Perrow, 1986a, 172).

Jackall writes, "because of the interlocking character of the commitment system, a CEO carries enormous influence in his corporation. (…) His word is law; even the CEO's wishes and whims are taken as commands by close subordinates on the corporate staff, who turn them into policies and directives" (Jackall, 2010, 23). So, despite limitations in the power of executives and top managers, the *Post NA* discourse recognises that they remain in many cases the most powerful actors of organisations. Most of the time, their power is much greater than that of any other actor of their organisation even if there are restrictions.

Indeed, their strategic choices constitute and remain the daily constraints that many other actors have to deal with in order to perform their tasks: "bureaucracy expands the freedom of those on top precisely by giving them the power to restrict the power of those beneath" (Jackall, 2010, 93).

Framing Strategy

So, an understanding of strategy requires use of the lenses of the linear, adaptive and interpretive schools, one of which may be called upon more than the others depending on the specific cases under investigation as well as our personal intellectual sensitivities and purpose (it is the interpretive one which is favoured in *Post NA*). This classification provides an interesting first approach because a central concern for organisations and their top managers is indeed to opt for the adaptations and orientations, which will bring success to their organisation in its environment, whether these adaptations are incremental or top down. These adaptations include the ability to successfully align position, choices and action (Johnson, Whittington, & Scholes, 2013).

Strategic choices include product and service developments, their price, diversification and geographical implementation. These can be promoted through internal innovation, merger and acquisition, alliances or joint ventures. These choices are always in relation to a picture of the position of a company by its top actors, and issues related to implementation are, of course, paramount to any success. Indeed, these strategic moves must be in line with the capabilities of an organisation, which are anchored in its history, people, culture and assets and turned into choices of organisational structure (e.g. centralisation vs decentralisation) and employees, along with symbolic (e.g. discursive, cultural) and motivational (e.g. incentives, career evolution, workplace and job satisfaction) aspects, while satisfying a diversity of shareholders and stakeholders.

In this context, a synthetic approach presented as an embryo of a general theory of strategic management is provided in the framework of dynamic capabilities, which can be decomposed into three main properties: sensing, seizing and transforming (Teece, 2007). Of course, not all business domains face similar complexities, but in the context of globalised markets with their fast pace of innovation, competition and evolving digital and financial environments combined with social change (these points are developed in Chapter 4), translated in a range of varying expectations and consumer trends, orienting organisations successfully is without a doubt a very complex task (Eisenhardt, 2002; Teece, Peteraf & Leih, 2016). It reveals the full complexity of social realities for which a diversity of actors participates in shaping top decision-making processes. In this respect, books and articles on strategic failures are quite useful for providing relevant insights, which one can easily link to the outcomes of research on errors very briefly introduced earlier.

What We Know about Strategic Failures

Analysing Strategic Failures

Although the strategic literature favours recipes for success and stories of success over those of failure (Pfeffer & Sutton, 2006), a certain number of books and articles have nevertheless been published in the past 15 years on this subject, offering systematic treatment grounded in case studies (e.g. Finkelstein, 2003; Sayles & Smith, 2006; Hamilton & Micklewaith, 2006; Carroll & Chunka, 2008; Kerdellant, 2016). Despite the absence of theory of strategic mistakes, error or failure, as Kerdellant asserts (Kerdellant, 2016; see also Shimizu & Hitt, 2011), books are organised along similar lines.

They are based on examples grouped into categories of strategic failures, followed by psychological, cognitive, organisational and sociological explanations of these failures, targeting both individual (CEO) and group levels (top management team), sometimes from an institutional or political economy viewpoint (e.g. Sayles & Smith, 2006). The selected categories are consistent with the description found in the literature as indicated earlier.

They thus describe strategic orientations that turned badly, as for instance in the case of a merger, acquisition or product launch. There are plenty of case studies available. One example of a product launch disaster that is often analysed because of the billions of dollars lost is the failed innovation "Iridium" by Motorola in the phone industry. In general, these failures are grouped into a recurring pattern of executives and top management teams' inability to perceive the problem associated with their strategy.

This idea is highly consistent with the interpretive view described earlier (Table 3.1). Indeed, a business strategy can turn into a sort of paradigm or mindset shared by executives and top management, which then becomes a deeply held ideology, resisting against warnings of problems.

The underlying cognitive, psychological and sociological reasons for these failures include the classical biases than one finds in the literature such as optimistic but also confirmation biases, and from a group perspective, the issue of a lack of a devil's advocate position to counteract the phenomenon of groupthink. As Finkelstein writes, "in a world of check and balances, when there is no real countervailing force to a CEO, individual preferences can dominate" (Finkelstein, 2003, 42).

This point indicates not only the importance of how top management teams are structured and able to decide on the basis of potentially conflicting views, considering uncertainties involved in any strategic moves, but also importantly, the necessity of a sufficient balance of power to divert from wrong paths when the CEO has his or her mindset on something.

Although not found in these fairly recent books on strategic failure, there are concepts available in the strategic literature that would provide appropriate analytical lenses for understanding these events. Let's mention, without being exhaustive, three such notions derived from a process perspective of strategy included in the interpretive view (Table 3.1): "strategic drift" (Johnson, 1987), "strategic dissonance" (Burgelman & Grove, 1996) and "escalation of commitment" (Staw, 1997). Table 3.2 offers a short description of each concept and its relevance.

Escalation of commitment is highly relevant here and defined as situations "where losses have been suffered, where there is an opportunity to persist or withdraw, and where the consequences of these actions are uncertain" (Staw, 1997, 192). This definition fits quite well with the view of Finkelstein, who argues through his analysis of numerous strategic failures that "the real causes of nearly every major business breakdown are the things that put a company on the wrong course and keep it there" (Finkelstein, 2003, 138).

Reasons for keeping the wrong course in the case of escalating commitment are diverse, individual and collective, ranging from the nature of the project involved (e.g. size, expected payoff or availability of possible alternatives) to psychological (e.g. illusion of control, optimism, self-justification) and sociological (e.g. social justification, leadership norms) dimensions (Table 3.3).

One strong message from this literature is that strategic failure is not to be too simplified retrospectively. "Decision makers must accept that the tendency toward errors is deeply ingrained and adopt explicit mechanisms to counter those tendencies" and that "the really aware executives realize the limitations they face. So they redouble their efforts, insisting on greater vigilance and deeper analysis" (Carroll & Chunka, 2008, 197).[3]

TABLE 3.2

Relevant Concepts for Understanding Strategic Failure

Concept	Short Description	Relevance for Strategic Failure
Strategic drift (Johnson, 1987)	Incremental loss of fit, despite strategic adaptations, between the company's products (or services) and consumers' choices	Degree of strategic failure depends on the extent of uncoupling between products (or services) and consumption of these products by customers
Strategic dissonance (Burgelman & Grove, 1996)	Process triggered by leaders' appreciation that products (or services) need to be questioned in relation to market changes	Absence of dissonance potentially leads to failure because need for change is not perceived or happens too late
Escalation of commitment (Staw, 1997)	Situation of strategic uncertainty over existing problems and whether or not (and to what extent) to maintain a course of action in relation to these problems	Strategic failure consists in choosing to maintain a course of action instead of diverting it

[3] Authors conceptualising successful strategic decision-making processes specify the kind of interactions within top management teams along these lines (Doz & Kosonen, 2010; Eisenhardt, 2013). It covers dimensions of conflict resolution, data sharing and updating among executives who are used to working together and also knowledgeable in their industry.

TABLE 3.3

Escalation of Commitment

Escalation of commitment	Nature of project involved	Size
		Expected pay off
		Availability of possible alternatives
	Psychological dimensions	Illusion of control
		Optimism
		Self-justification
	Sociological dimensions	Leadership norm
		Social justification

Source: Based on Staw (1997).

Therefore, following his investigation of strategic failures based on retrospective interviews with executives, Finkelstein warns against the simplistic explanations commonly formulated in hindsight. Reflections such as "the executives were stupid" or its opposite "executives couldn't have seen what was coming" can be heard. Interpretations that "it was a failure to execute" or that "the executives weren't trying hard enough" can be found too.

As indicated earlier, Starbuck suggested a very similar cautious attitude about any retrospective appraisal. But Finkelstein goes further, adding and contending that "the personal qualities that make this awesome scale of destruction possible are all the more fascinating because they are regularly found in conjunction with truly admirable qualities (...) most of the great destroyers of value are people of unusual intelligence and remarkable talent" (Finkelstein, 2003, 213).

Degree of Strategic Failure

Following this summary and in generic terms, strategic failures can be described as the inability to provide a suitable orientation and operating mode for a company in evolving (and very often tough) business environments. This definition creates the fundamental tension for any executive in a complex world as indicated by Starbuck, which is found as a recurring theme in the literature intrinsically associating strategy with risk.

"Risk is inherent in business. If you never took risks, you'd never chart new ground. You would never revolutionize a market or a product or an industry" (Finkelstein, 2003, 271). One could add, with Kerdellant, however, that there are degrees of risk taking and strategic failures. This author distinguishes between a broad range of cases and implies a kind of scale in the structure of the book, although without conceptualising it further. Based on this structure, I suggest a distinction between strategic mistake, strategic failure and strategic fiasco.

Examples of strategic mistakes abound when products or services fail to meet their customers' expectations, when top managers miss an important trend in markets, when mergers or acquisitions do not lead to the expected gains, or also, when geographic implantation of products or services in new countries misses the local cultural scripts that would allow their local success.

These remain mistakes when the survival of the company is not involved. These mistakes can or should be part of not only expected business complexities but also the learning processes of managers and top managers. But when executives and top managers make highly risky moves threatening the entire assets of the company, the notion of strategic failure becomes more appropriate (note that the reverse is true too: when executives fail to anticipate and adapt to major changes in markets, they also threaten the survival of the company).

And, when large-scale fraud produced by executive hubris turned into unethical practices and maintained in specific favourable institutional contexts ends in scandal, we move into the next stage of strategic breakdown, something that could be called strategic fiasco. Perrow suggested in this situation to use the qualification of executive malfeasance. The notion of "rogue executive" has also been proposed for this form of behaviour (Sayles & Smith, 2006), following a series of scandals including Enron in 2001, and a bit later, the subprime crisis in 2007/2008 in the poorly regulated financial capitalism (also described as predatory capitalism, see Chapter 4).

Defining and Summarising Strategy in the Context of *Post Normal Accident*

To summarise the above, one can retain that to approach and understand strategy, its complexity, ambiguity and uncertainty, one needs to describe and consider (see Table 3.4), first, the main tasks and activities associated with strategic decision-making (e.g. product diversification, acquisitions, organisational design) and, second, the complex social business environment populated by a diversity of actors (e.g. boards of directors, top management team members, regulators, media, consultants).

Third, to explore strategy, one can rely on various analytical perspectives (i.e. linear, adaptive, interpretive) and, fourth, on different concepts (e.g. strategic dissonance, strategic drift, escalation of commitment) shaped by cognitive, psychological and social realities that help explain, fifth, the mechanism of strategic breakdowns of various scales (i.e. mistake, failure or fiasco).

A definition of strategy could be the "choices made by top managers over the main orientation of a business in its market and the ability of these orientations to be successful considering the human and technical capabilities involved". To the question of "How is it that top managers are likely to make mistakes or errors when it comes to designing, choosing and selecting options which considerably shape operational context of high-risk systems?" asked earlier in this chapter, a beginning of answer with the help of strategy studies is being shaped.

TABLE 3.4

Key Aspects of Strategy in the Context of *Post NA*

Strategic tasks and activities	Merger, acquisitions and ventures, profits' targets, cost-cutting, downsizing, budget allocations, choice of organisational structure, choices of geographic or product (services) diversification, choice of top personnel, design of work methods, motivation and rewards systems, management and conflicts resolution style
Complex social business environment	Boards of directors, top management team members, regulators, shareholders, auditors, non-governmental organisations, unions, medias, consultants, competitors
Analytical perspectives on strategy	Linear (decision then implementation), adaptive (incremental and emergent) or interpretive (cognitive, cultural and power view of strategic decision making)
Strategic decision-making failure concepts	Strategic drift, strategic dissonance, escalation of commitment (in relation to cognitive, psychological and sociological phenomena featuring structural, cultural, power and historical aspects of strategic decision-making processes)
Degree of strategic breakdowns	Mistakes (e.g. product not meeting its customers), failures (e.g. unsuccessful mergers and acquisition) and fiascos (e.g. unethical behaviour, fraud, scandals)

With this in mind, the connection with the domain of disasters and safety should obviously be stronger, and this link needs to be established. Two combined options are now proposed for this purpose. The first one is to conceptualise safety in the context of companies' strategy more explicitly than available so far. As introduced earlier, top management decides about key moves such as acquisitions, products and geographic diversification and profits' targets but also structure of organisation, style of management and key personnel. All of this strongly contributes to specific relations and interactions between a diversity of actors in a company. A proposition is therefore to consider, as a prerequisite, that safety is produced in the context of specific strategies that should be made more visible.

Considering safety this way adds a dynamic approach, something that is missing in many research traditions, as contended by Griffin, Cordery and Soo (2015) with the help of the dynamic capabilities framework of Teece (2007). "There is currently no conceptual framework for describing qualitatively different kind of change in safety systems (…)" (Griffin, Cordery & Soo, 2015, 2). As discussed, companies evolve constantly, and changes are strongly correlated to strategic moves. In the literature on strategic breakdown, key moments in the life of an organisation such as a merger, the launch of a new product, internationalisation, new projects or organisational change are always critical.

These moves can succeed or fail, in isolation or all together. These moments also include changes in structure of companies or change of key managers placed in new positions, in relation to these changes, creating new kind of interactions between the diversity of actors of organisations. From a safety point of view, these analyses are important because they reveal a quite central feature of organisation life. These are moments when top management is deeply involved in the process of making sense of the uncertainties of how well their strategic decisions do.

This leads to the second option to be combined with the first. Strategy and strategic failures are cognitive, psychological and sociological realities as discussed earlier. It is a product of executives and top managers' decision-making processes in their complex work, social and business environments, as well as wider institutional context. This second option refers to this aspect of strategy that points to the type of relationship and interactions within top management teams and beyond, within and outside organisations, and the known phenomena ranging from biases to group think that can contribute to miss signals. These phenomena influence deeply what has been described as strategic dissonance, strategic drift or escalation of commitment (Table 3.5).

Illustrating Safety as Strategy

Let's illustrate these options of safety as strategy with three different cases. These cases can be differentiated along two dimensions: company's size and magnitude of event. The first case (called *silo case*) concerns a relatively small

TABLE 3.5

Strategic Concepts Applied to Safety

Concept	Short Description	Relevance When Translated to Safety
Strategic drift	Incremental loss of match, despite strategic adaptations, between products (or services) of the company and consumers' choices because core strategic paradigm resists	Incremental moves can make the company drift away from safe practices without being part explicitly enough of the picture of top leaders
Strategic dissonance	Process triggered by leaders' appreciation that products (or services) need to be questioned in relation to market changes	The absence of dissonance between safety problems and strategy create the possibility that course of actions remains unchanged despite its potential consequences
Escalation of commitment	Situation of strategic uncertainty over existing problems and whether or not to maintain course of action in relation to these problems	Incidents and accidents can be considered as problems in need of reappraisal of strategic orientations (or implementation), but actions taken fail to tackle the problem adequately

and national company in the business of food grain storage and distribution (250 people) and a fire without any casualties but with product (grains) and infrastructure damages. The second one (called *pyro case,* for pyrotechnic) involves a bigger company (1400 employees) producing and selling dynamite, operating in several countries and an explosion causing the death of four employees.

Finally, the last one is about a multinational present across the world (100,000 employees) and a series of disasters causing environment damages and many casualties between 2005 and 2010 (called *petro case*). These three cases increase successively the size of organisation involved and magnitude of impacts of events. They provide enough material to address the question of strategy as framed in this chapter. The aim is to show that the three can be described as strategic breakdowns of various degree, from mistakes to fiasco. The first two are first-hand personal empirical studies with access to top managers and executives of the companies (Le Coze, 2010, 2012).

The third one is the BP case which relies on secondary data but has been extensively covered by a series of reports and books over the past 10 years (e.g. Hopkins, 2008, 2012; Bergin, 2012; Lustgarten, 2012; Le Coze, 2016). These cases are presented each in turn briefly to highlight the specific aspects of importance for *Post NA* discourse. They do not go into the details of the technical or human factor aspects but focus on the relationships between events, organisation and strategy.

Note that these missing levels of descriptions about engineering, human factors and organisation and regulation do not mean that these are not available. In fact, these illustrations depend on in-depth ethnographic type of investigations which can only provide the level of details needed to judge a specific situation from a strategic angle. They, however, slip in the background here because the focus is on strategic level of understanding and interpretation. But it is clear that methodologically, this type of analysis and appreciation can only be inferred when a rich empirical material of a diversity of artefacts and activities of operators, engineers and managers from top floor to top management and regulation is available, with the help of the lenses of several research traditions.

One word about main differences between these three empirical studies, beyond technology, in relation to their size. It stands to reason that one cannot address strategy of a small and local company as one of a multinational. Because of their complexity, multinationals face specific challenges unmet in smaller-scale organisations (Morgan, 2006). However, despite this obvious difference, an approach of safety as strategy can apply equally well. And there is a heuristic value in comparing them even if from the point of view of the design of a research agenda and its methodological implications, studying safety from a strategic angle in a small/medium-size company and a multinational entails very different prerequisites and complexities.

Silo Case

In the silo case, a fire occurs in one of the silos of a company without casualty but with costly damages. The investigation reveals that the silo was not well managed by a young recruit. This young employee should have monitored heat parameters according to a determined frequency but was not in a position to perform his work properly because of lack of training, experience, adequate working conditions and supervision. Risks associated with silo, from a major hazard point of view, are fires or explosions. It is imperative to check temperatures of storage. This constituted a real drift from expected practices.

The executive manager translated this event in the following sentence, as resulting from "a faulty employee badly supervised". The situation could be framed in a different way. The context of the event was indeed embedded in a change of structure, of key managers and of relationships and interactions between managers created by this new executive manager as part of his strategy for the company, creating new operating constraints. Let's develop this description.

This new executive manager replaced two previous executive managers who both had operational experiences. Industrial activities of this organisation include about 70 silos that, depending on their size and complexity, are operated by one or several operators. These units are spread on a geographic area and grouped into supervised entities managed by appointed people. Without operational experience, this newly appointed executive manager wished to distribute the operational side of his activity to operational managers who would report directly to him.

The idea was for him to concentrate on key moves for the future of the organisation, such as the prospect of merging with other companies, and on the administrative dimension of his position. The company was in no specific economic difficulties and had a good position in its market, based on the storage and distribution of crops in this area of the country. One trend was, however, for this type of business to grow through combining resources with neighbour organisations, basically extending geographic cover, and mutualising activities. The associated change of principles of organisation and managers was directly linked to this strategy.

He recruited, with the help of his human resources manager and an external consultant, three profiles inside the company to take on the new positions of operational managers who would report directly to him. These new positions represented a challenge because it meant a higher number of silos and more people to supervise (with a younger workforce because of demography dynamic), hence the need to create also new positions at an intermediary level between operators and these operational managers.

This new structure was reinforced by a formal reporting system designed to regularly provide updates on operational status of activities in a homogeneous way by the three appointed operational managers. Audits by the quality, health, safety and environment (qhse) department were also to be performed on top of this. A problem was that one of the new employees in the position of operational manager to report directly to the new executive came from the base of the organisation, with no previous experience of managing several silos and people. Without knowledge at this level of management, he had much to learn, but no specific training support was provided.

The new executive and human resource manager thought highly of him, and when a certain number of problems cumulated because the new manager struggled in the new position, they remained unacknowledged by the executive, and this new operational manager hid rather than openly discussed his own difficulties due to his lack of experience. The two newly appointed managers out of the three noticed the attitude of their colleague and did not necessarily like his management style, but they did not intervene because they lacked knowledge of how exactly the situation was from an operational point of view in his area.

The qhse manager complained about problems in the way safety was handled under the supervision of this new manager, but the executive did not respond positively and left the situation to a status quo. What happened is that the change of executive, organisation and managers had created a different position of authority for the qhse manager. Theoretically, her position was the same in the organisation structure, but the new executive favoured the opinions of the new operational manager who evolved in his new position and who the new executive wished to support in this challenge (without, however, much training support).

Following the investigation of the fire from a human and organisation angle, the new executive finally realised that a certain number of problems existed in relation to the difficulties met by the new manager (which led to a neglect of the working conditions and practices of the young recruit on the burned silo) and that changes had to be made to his own style of management, choice of managers, or in other words, his own approach of dealing with his strategy. Instead of "a faulty employee badly supervised", it was therefore more appropriate to talk of "a new strategy with unanticipated and unmanaged weaknesses". Two different levels of explanations are not exclusive, but the latter constitutes the context of the former.

Pyro Case

In the pyro case, an explosion of one of the production cells of a plant producing dynamite killed four persons, including three who shouldn't have been there at the moment of the explosion, questioning the management of the site that morning, in particular circulating paths and coactivity in the plant. As for the silo case, the context of this event was a series of transformations in the way the plant was managed and handled from the top over the years. But the complexity of this organisation is higher than in the silo case because of multiple plants (including some abroad) with many more employees than in the case of operating silos and a higher number of employees globally with more differentiation of expertise, roles and positions within the group.

Leader on its national market and part of the top four worldwide, the company had decided to extend its activity to include an engineering service on the use of dynamite along with its core business of producing dynamite cartridges. Family owned and without any debts, the financial background of the company was favourable to many investments in design, which were made over the years in order to improve safety, health and productivity all together in the plants.

However, as part of the evolution of the company under the strategy, some changes were made to the organisation in terms of technology, key managers and their activities. These constituted the background of the accident and its casualties. A certain number of these strategic changes were decided at the level of both plant and corporate management. Overall, these organisational decisions created operational problems. First, a new plant manager without operational experience was appointed.

Second, this new manager, after a short period of time, divided its time between two plants in order to optimise their management, in particular when it came to dealing with social climate and to capitalising the expertise needed to deal with complex environmental and safety regulatory requirements. One consequence in this particular case was that power shifted to the production manager of the plant where the accident occurred. Without much experience, the newly appointed manager relied on the more experienced production manager, which created a new trade-off situation in the factory, at the expense of the qhse manager.

Third, and combined with the previous strategic changes just described, was the appointment of a new technical expert at the corporate level. This modification (which was just one of employee and not of function) contributed nevertheless to alter further the power of the safety department in the same plant. The qhse manager relied previously on this individual at corporate levels, who he knew well, to get his message across about safety issues at this level of decision-making. In his absence, the qhse manager lost his influence.

Combined, these strategic choices of organisation (e.g. appointing an inexperience manager, dividing its activity between two factories, replacing the technical expert at top level) created new social interactions, removed the site manager from daily practices in the factory, increased power of the production manager and weakened the ability of the qhse manager to play his operational role.

For instance, the days before the accident, it was decided to reopen some production lines in the plant in order to deal with a situation of an excess of product to be processed, and coming from another factory of the company, abroad. This new and temporary operational constraint had the effect of causing problems of circulation in the plant and contributed to the number of casualties following the explosion because it increased coactivities and their associated complexities, without establishing a management of change process which could have anticipated such issues. This was an example of failing to provide a favourable balance between safety and production in the absence of a site manager permanently in the factory.

Without the possibility to confront and to oppose to this option at top management level, it was not possible for the qhse manager to play a stronger role. These new social configurations of relationships between managers and the information flow it created throughout the organisation from plant to top management level following this strategic move in the organisation and employees' appointments remained invisible to top management until the accident revealed the many associated operational problems. The event shook their beliefs that the technological investments made over the years for designing a safer plant were sufficient to determine how safe the factory practices were and that this social dimension of safety counted as much as its technological side.

Petro Case

The petro case (BP) cumulates, in a period of 5 years between 2005 and 2010, three disasters in three different branches of this company's activities (pipeline oil transport, refining, offshore exploration), challenging the management of process safety across the company. In the first disaster, a poorly maintained refinery (Texas City, 2005) caused an explosion and the death of 15 subcontractors. In the second, an oil spill (in Prudhoe Bay, 2006) was created by corrosion because of a lack of inspection and maintenance programs. And in the last case, a deep water well (Macondo) failed to be secured and leaked gas on the platform (Deepwater Horizon, 2006) that exploded, creating an environmental disaster and 15 casualties.

With this case, another layer of complexity is added in comparison with the silo and pyro cases because of the bigger scale of the multinational involved. However, because of the wealth of material available (investigation reports, books, articles), it is possible to approach this story also from a strategic angle as done for the two previous examples. What can be inferred indeed from the number of analysis and interpretations on this case study is that the firm suffered this series of disasters as a result of strategic moves.

The story of this multinational is one of the organisational transformations in line with a strategy from the middle of the 1990s onwards, which had clear detrimental consequences for process safety, creating the conditions for these three major events to occur. And these transformations are closely linked to the strategic orientations developed by the executive and his team throughout a decade, including several important orientations in terms of diversification, acquisition and structural changes of the organisation.

One key initiative consisted in finding new petrol fields for the company, requiring an expertise in dealing with multiple political, social and technological challenges. The strategy also implied buying and merging with other petroleum companies in order to grow in size and to prevent the risk of aggressive takeovers by other petroleum multinationals at a time when the company was smaller in size.

From the organisation point of view, the old vertical structure was progressively transformed into a networked configuration in which the internal (safety) engineering expertise was decentralised and business units (BUs) were created with a higher degree of autonomy across the branches, starting from exploration and then to refining. This was intended to increase decision-making speed and flexibility at the regional/branch levels. Along with these strategic orientations, a cost-cutting policy was activated, expecting financial targets to be met by BU managers.

Diversification, merger and acquisition, restructuring and cost cutting in a period of low petrol price characterised this financially and commercially successful strategy. However, the disasters challenged the ability of the multinational to maintain safe operations. The networked configuration had created an unfavourable context for safety trade-offs at least in some parts of the multinational, under cost-cutting pressures. Favouring decentralised mode of operating for more autonomy and fast decision-making processes at the level of branches and BU altered, in a context of strong cost-cutting imperatives following the merger and acquisition, the check and balance principle expected of high-risk systems, considering their hazardous potentials.

Safety as Strategy

Making Sense of the Three Cases from a Strategic Angle

In the three cases, it is highly relevant to make sense of the events as products of strategical choices, which materialised in operational constraints and structures affecting peoples' interactions and safety practices. Causality of accidents and disasters cannot be inferred without a clear conceptualisation of a company's strategy. Strategy establishes, creates and produces the specific operating constraints within high-risk systems. What is clear is that strategic changes constitute a key topic of analysis in order to contextualise unsafe practices whether, when operating a silo, a pyrotechnic plant or an offshore platform and whether in a multinational- or a national-based company.

Despite their difference in both organisation size and magnitude of events involved, each can be translated as strategic flaws of various degree, something implying to reveal the psychological and sociological realities of top decision-makers (as introduced earlier when analysing studies of strategic failure). Because safety is one parameter among others in companies, an adequate strategy is one that establishes a complex balance between other goals including productivity, return on investments, social climate, human resources and any other legal or regulatory expectations in relation to work, environment or commercial contracts.

This implies potentially that executives or top managers can be reluctant to change their course of actions derived from their strategies until problems are obvious or visible enough so that they must accept to alter and to modify the mindsets associated with these strategies. With these three examples, we see indeed how a strategy of executives and top management teams turned into a mindset associated with operational constraints, an organisational structure, positions allocated to managers, management styles and social patterns of interactions all of them creating specific safety conditions. One interpretation is indeed that strategic orientations, as shown earlier, provide a background from which treatment of signals of safety problems to be dealt with is framed.

The literature on strategic mistake or failure as introduced in the previous section, combined with conceptual insights (e.g. strategic drift, escalation of commitment), provides much support to legitimise such a line psychocognitive view of explanation. Because of the strategic cognitive mindset of powerful actors in these organisations, it took an event (or several) of sufficient proportion to modify its associated perception of the problems. With regard to Farjoun and Starbuck's question "if an organization is striving to exceed some of its limits, some people are pushing for very extreme performances. Who is exerting such pressure, and for what reasons?" (Farjoun & Starbuck, 2007, 562), the three cases provide therefore some clues.

It is pretty obvious in the first case, the silo case, that the executive transformed the organisation to implement his strategy, that it took a serious event to question it and that the dynamic within the management committee and beyond failed to alert him about the potential consequences. The gap of interpretation of this executive expressed in personal interviews with him before and after the event showed that his beliefs had been shaken considerably and that what he thought to be an adequate approach proved to be quite weak in some areas.

The second case is very similar, but this time it was the idea that the design investments of the past years had made the plant a safer place, until the limits of the new organisational arrangements and relationship between managers combined with operational constraints proved also to hide some weaknesses and flaws in the management of the plant. For these two cases, silo and pyro, there was no deliberate cost-cutting strategy or deliberate trade-offs between safety and production but rather an incremental series of changes of the organisation in operational constraints and structure consistent with the strategy which created unexpected patterns from the point of view of executives and top managers.

The third case is more radical because of the extent with which the strategy clearly relied on strong cost-cutting policies across the company in the context of mergers and acquisitions associated with a new commercial philosophy emphasising financial achievements of autonomous BUs, at the expense of the strong centralised organisational configuration of the past. Pushing the logic of networks (decentralised structure) in a more centralised industry, this strategy was a much more ambitious, innovative and aggressive approach (strong operational constraints due to cost cutting) than the two previous cases.

The consequences were not obvious immediately and took a few years before creating the conditions for the highly visible events that followed and were maintained by a powerful and charismatic leader who designed his top committee to remain unchallenged. Thus, as for any strategic mistake or failure involving, for instance, a new product or a merger, "the real causes of nearly every major business breakdown are the things that put a company on the wrong course and keep it there" (Finkelstein, 2003, 138). In the three cases, changes derived from top managers' strategies that created new configurations affecting safety but resisted until problems could not possibly be seen as peripheral but as exposing instead some core weaknesses.

The events revealed the inadequacy of some key aspects of what the strategy consisted in. As described in the literature, processes such as these have strong affinities with notions such as "escalation of commitment" (Staw, 1997), "strategic dissonance" (Burgelman & Grove, 1996) or "strategic drift" (Johnson, 1987). These are useful concepts for safety as understood through a strategic angle of analysis (Table 3.5).

Strategic Mistake, Failure or Fiasco?

Should strategic mistake be discriminated from failure or fiasco in the safety domain as implied in the strategic literature and suggested earlier when exploring strategic breakdowns? Are there degree of safety strategic breakdowns? One intuitive option is that strategic mistakes could be associated with events without major consequences for the company (whether these consequences have material and human dimension, or social). These strategic mistakes would sound as warnings that something is wrong with a strategy. Strategic failures could be linked to high-consequences events.

This would indicate that a strategy has gone too far without realising its safety implications. But is this as simple? Is the degree of strategic breakdown linked with its consequences? This certainly needs each time a cautious treatment based on empirical data, but it seems indeed to make sense that the more a strategy challenges the ability to remain within the envelope of safe performances, the more likely it is to provoke large-consequences events.

Out of the three cases, the petro one clearly looks like a strategic failure, or perhaps even a fiasco. The extent of safety problems in the three events was quite critical, with strong warnings, for instance, from a refinery manager that resources were needed to improve integrity of installations.

Concealed in their strategic mindset of delegating autonomy to BUs, top management would not relieve the refinery from its cost-cutting objectives and constraints despite pleas to do otherwise by the site manager of the refinery where the disaster occurred. Why changing the course of action of a winning strategy?

In this case, the view of Carter, Clegg and Kronenberg strikes as particularly relevant for the petro case. "Strategy writes scripts for adulation today which can easily be rescripted as shaming and deceitful tomorrow. (...) the very thing that makes a company rich, powerful and successful, can make the same organization poor, its executive criminals, and its investor's debtors, as we have seen in countless corporate scandals. It is easy to get caught up in the success of a winning strategy and be blind to the corners it cuts".

Can the same be said about the two other cases? Although exhibiting similar patterns of powerful actors shaping the conditions under which trade-offs were operated across the company, the strategic analysis creates a different feel. A deliberate aggressive cost-cutting approach was not visible, and changes in the organisation were not as radical or innovative although clearly establishing new principles of decision-making processes and redistributing power within the organisation.

Still, the consequences of the pyro case with the four casualties and their reasons make it more of a strategic failure than the silo case, which could be classified as a strategic mistake. The strategic failure of the pyro case does not appear to be so much an aggressive cost-cutting approach combined with a radical innovation to structure of business as with the petro case. It seems more to be a case of the inability to perceive in time that safety is as much the product of its technological side than its social one, when dealing with highly sensitive and hazardous material (nitroglycerine). In relation to Farjoun and Starbuck's quotes earlier, which distinguished two types of explanations of organisation at and beyond the limits, Table 3.6 can be suggested.

TABLE 3.6

Safety Strategic Mistake, Failure or Fiasco

<table>
<tr><td>Strategic fiasco</td><td rowspan="3">"Organizations may be driven to try to exceed their capabilities by senior executives who pursue unrealistic goals because of insecurity, ambition, greed, hubris, jealousy or competitive zeal".
–
"Organizations may promise too much and overstep their capabilities, not so much as a result of conscious design but as response to cumulative flows of events or as unintended by-products of decisions and actions".</td><td>Petro case</td></tr>
<tr><td>Strategic failure</td><td>Pyro case</td></tr>
<tr><td>Strategic mistake</td><td>Silo case</td></tr>
</table>

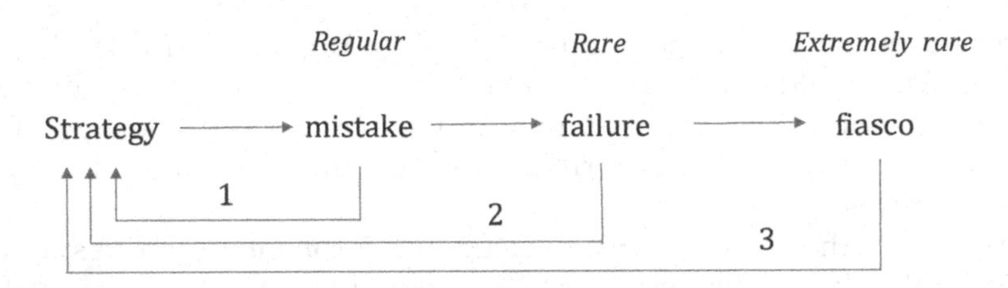

FIGURE 3.2
Learning loops between mistake, failure and fiasco in the safety context.

This distinction between mistakes, failures and fiasco implies indeed that a certain degree of safety achievement is correlated to strategy. A definition of strategy in the context of safety could therefore be the "choices made by top managers over the main orientation of a business in its market and the ability of these orientations to be successful without compromising safety considering the human and technical capabilities involved".

Considering that catastrophic events remain relatively rare for the majority of operating high-risk systems, one strong hypothesis is that, most of the time, strategic mistakes that materialise in incidents of low magnitude (and which are far more regular in industry) lead to a reappraisal of strategic aspects (e.g. operational constraints, organisational structure, key people and social interactions, management style) that explain their occurrence. This would mean that processes are most of the time in place to avoid moving from strategic mistake (1) to strategic failure (2), or worse, strategic fiasco (3) (Figure 3.2).

A New Hindsight Bias?

Retrospectively, this reformulation of safety as an outcome of strategic making processes meets the problem of hindsight bias or retrospective fallacy. These expressions target the comfortable position of observers who know the end of the story while people of the past face uncertainties and unwritten futures (see the earlier quote, "analysts have to surmount their tendencies to know more than they could have known", Hodgkinson & Starbuck, 2012).

Surely, any retrospective analysis of strategic mistake, failure or fiasco is a matter of judgement about the possibility or not for people to reasonably avert tragic paths, and in this chapter, strategic failure or fiasco. I believe that combining the strategic literature with safety research provides one of the most fruitful options to do this. To shift, or perhaps, best, to complement the analysis of human errors associated with front actors to strategic decision-making to question and to prevent potential errors from the top seems to be an important move.

Empirically, such a move can be translated in this sort of question: *how far can some strategic choices be openly discussed in organisations when it comes to their safety implications?* Such a question leads to consider cognitive, psychological and sociological dimensions of top management (Edmondson & Verdin, 2018;

Edmondson, 2019). This of course could prove to be a challenging empirical task as much retrospectively than in daily operations, to be detached as far as possible from legal and moral perspectives of top managers' activities.

Problem of Strategy or of Strategy Implementation?

First, signals can be ambiguous and the efficiency of modifications to course of actions too (e.g. escalation of commitment); this is the remark of Farjoun and Starbuck: "people and organizations do not always know how far they are from the true limits or the extent to which limits are elastic, relative, or arbitrary. Therefore, progress in general, and exceeding limits in particular entails ambiguity, risk and uncertainty" (Farjoun & Starbuck, 2007, 543).

Second, there are plenty of opportunities to shift the blame on other actors of the company, on technology or on market constraints when problems arise. When shown retrospectively that managers below top management teams did not meet expectations, it is tempting indeed for executives and top managers to consider that they should have performed better, in order to shift the blame at lower levels.

For executives and top managers, it is more comfortable to consider events to be problems of strategy implementation rather than of their strategy. It is perhaps in some cases rightly so, but sometimes, it is not. This situation was best expressed by this executive of the silo case who described the fire as a result of "a poorly performing operator badly supervised" (namely a problem of implementation) excluding the context of his own strategic choices which created specific conditions explaining the event better than his restricted and local focus.

The same could be said about the two other cases, pyro and petro, even more so that they are bigger organisations with many more possibilities of challenging managers and middle managers on what they were supposed or expected to do, namely finding ways of not sacrificing safety over production. It was the case of this BP Texas City refinery manager who was denied more resources to improve safety in the plant despite an appalling situation.

There is a need in this respect to refine the analytical categories associated with strategy between several levels of decision-making processes such as corporate and BU or site levels depending on size of companies. In particular, in multinationals, one can find a very high number of these levels, with possibly different implementation of strategic orientations of senior managers. There exists a possibility of decoupling between corporate activities and operational ones in financialised contexts, as shown, for instance, in studies of multinationals (Kristensen & Zeiltin, 2005). What degree of autonomy is left at the level of sites? How much of what is happening in sites travel to top management levels to inform about problems reflecting on strategic constraints? How much is kept at BU or site levels?

Thus, it is not unheard-of cases for which "managers at the middle and upper middle levels are often left to sort out extremely complicated questions about technology, investment, and their bosses' desires and intentions"

(Jackall, 2010, 85). Executive of course never explicitly asks for making decisions that would be detrimental to process safety. However, the context that they create, and the rewards associated with meeting targets despite tough operational constraints, can imply such decisions as much as explicit oral statements would, indeed "pushing details down to protect the privilege of authority to declare that a mistake has been made" (Jackall, 2010, 21).

An example is provided by Hopkins analysing the BP Texas City accident (2008). "BP's cost challenges had this characteristic: senior executives demanded cost cuts and left it to others further down the hierarchy to ensure that these cuts were not at the expense of safety. Lower-managers responded as best they could to these conflicting requirements, but, inevitably, safety was compromised" (Hopkins, 2008, 81). It is always possible in the absence of such statements by executives to pretend that they did not intend to do so and that such a rationale for decision-making was never implied. They are indeed happy to get credit for their successes but not for their failures that they tend to attribute to others or to external circumstances (Baumard & Starbuck, 2005; Starbuck, 2009; Quinlan, 2015).

A New Reductionism?

The hindsight bias (or retrospective fallacy) is not the only problem associated with an emphasis on strategy, and there is also the risk of creating a new reductionist account of events. Simplifying accidents to front-line human errors was rightly criticised in its time, so a focus on strategy might also be criticised to be a simplification of far more complex dynamics. Indeed, have Perrow, Hopkins or Vaughan not insisted on the complex, multidimensional and sociotechnical nature of safety and disasters instead? Are technology, structure, goal and environment not analytical dimensions to be articulated precisely to avoid any simplistic accounts as explained at length in Chapters 1 and 2?

One answer to these concerns is that any sociotechnological analysis of disaster and safety relies on an explicit or implicit attribution of relational causality between multiple dimensions such as technology and task, structure, culture, strategy and environment. To argue for the importance of strategy is not do deny the causal relevance of the other dimensions, it is to consider strategy pivotal to the causal relationships between all of these dimensions. Let's comment this assertion. Strategy is not determined but constrained by specific regulatory environments. Organisational structure is determined by strategical choices. Culture is constrained, influenced or shaped by strategy of top management.

Technology can constrain strategy but does not determine it, and tasks are shaped by strategical choices. It is not to say that strategy is unconstrained and unrestrained, but that strategy is the prism through which one unravels the causal interplay between these dimensions. As Farjoun and Starbuck write, "of course, since organizations choose their environments and technologies, the issues is partly one of business strategy" (Farjoun &

Starbuck, 2007, 553). The fact that there is a diversity of distinct companies operating similar technologies, in similar markets and similar regulatory environments, is explained for a nonnegligible part by the history of strategic moves over the years in these different companies.

Highlighting strategy is therefore not a new reductionism but an emphasis that helps articulate together other dimensions such as technology (task), structure, culture and environment. Focusing on strategy is not a new simplification, it is needed to understand safety in daily operation of high-risk systems and to make sense retrospectively of disasters. Note also that an emphasis on strategy is not a reductionism also because it must be built empirically and conceptually through interdisciplinary work, namely based on the use of different research traditions (Le Coze, 2019a). Strategy in Figure 3.3 replaces goal and is in bold to stress its centrality in case studies.

For this reason, empirical studies of daily operations in high-risk systems must consider the influence of powerful decision-makers. These can be site managers or corporate executives. As illustrated with the silo, pyro and petro cases, work situations and practices of a multiplicity of individuals in organisations are constrained by design choices made by top management at site, branch or corporate levels, for instance, the choice of organisation structure, the appointment of key managers at various positions in board of directors or management committee, the composition and number of operators in production teams and also, of course, of great significance in the context of safety, operational constraints linked to strategic moves such as merger and acquisition or business reorientation including investments and disinvestments in assets and profits' targets.

It is therefore difficult not to associate safety with these choices, and to understand how they are made, by whom, under which circumstances, and most importantly how and why they resist to signals that the consequences of these choices are potentially detrimental to safety. This of course is a very important issue for the study of daily operations, which is currently poorly investigated as the focus is today predominantly on shop floor analysis or

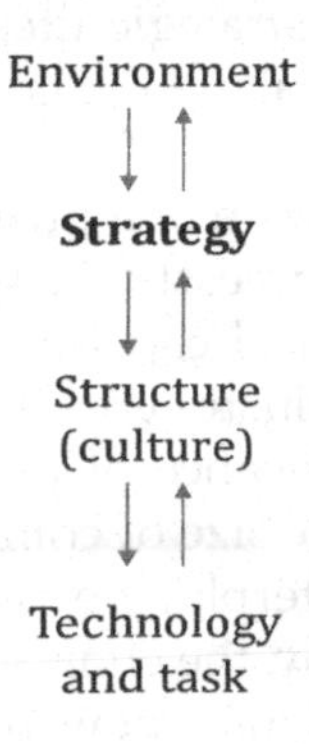

FIGURE 3.3
Strategy as a core dimension.

organisations but without introducing explicitly strategic level of analysis in many studies, a point also made by Hayes. "Despite decades of safety culture research that emphasises the role of safety leadership, little normal operations research addresses the practices of senior management" (Hayes, 2019, 199). The lack of description of top management in relation to safety through strategy can lead to a neglect of this dimension as the basic ingredient of safety.

Summary of Chapter 3

Perrow and Hopkins have both stressed how important it is to understand disasters through a consideration of the top of organisations within their environments. However, it is argued that they have done so only partially, without explicitly connecting this corresponding level of analysis with the strategic literature. Starbuck along with Farjoun has done so more explicitly but without pursuing this time their proposition empirically within a specific agenda that would connect safety and strategy.

Yet, strategy is one key dimension of the daily operations of high-risk systems because any working situation reflects somehow choices made by top decision-makers at different levels of organisations, depending on their sizes and modes of operating. Companies are steered through strategic orientations, and their success and failure are the products of powerful top decision-makers who decide or not to avert or to modify certain choices made at one point or another.

Strategic breakdowns have indeed been studied in business contexts but never applied or translated into the field of safety, although such a translation appears helpful. A proposition is to graduate the degree of strategic breakdown through three levels of intensity: strategic mistake, failure and fiasco. Moving up the notion of human error to the top of companies offers a much-needed refinement to characterise high-risk systems operations, borrowing for this purpose in the strategic literature notions such as strategic drift, strategic dissonance or escalation of commitment.

A definition of strategy applied to safety is then the "choices made by top managers over the main orientation of a business in its market and the ability of these orientations to be successful without compromising safety considering the human and technical capabilities involved". This proposition to frame safety as strategy is illustrated with three case studies (silo, pyro and petro) which show the relevance of proceeding with an emphasis on this dimension, regardless of the size of companies. Strategy is pivotal to an understanding of the causal interplay between technology, task, structure, culture and environment. Finally, the chapter stresses the need to limit both the hindsight bias and the risk of a new reductionism that would isolate strategy from its complex decision-making contexts, to which we now turn.

4

From Component to Network Failure Accidents

Introduction

Normal Accident (*NA*) had a broad scope. Moving from the first to the second thesis as explained in Chapter 1, Perrow's framing of high-risk systems expended to articulate together technology (task), structure, goal and environment. Perrow saw organisations as shaping and being shaped by society. With this move, the notion of component failure accident was additionally developed by Perrow to characterise disasters which happened despite knowing that they could be prevented in principle.

Component accidents are due to inappropriate goals of organisations in weak (regulatory) environments, which allow for these component accidents to occur when production is favoured over safety. Following this idea, and translated in the more contemporary term of strategy, Chapter 3 shifted the focus from front-line errors to errors from the top arguing further that accidents should be framed as strategic breakdowns.

Chapter 2 showed that Hopkins concurs and follows Perrow's interpretation while showing a more practical than critical side by elaborating a normative model appealing to readers wishing to think about improving their practices. He also convincingly persuades his readers of the idea that disasters could be prevented, if only appropriate practices had been implemented, which leads us again to top decision-makers.

But he introduces something else that Perrow did not because Perrow wrote *NA* at a different time, in the 1970s and 1980s. Hopkins, writing in the 1990s onwards, captures in his work one of the profound transformations of the past three decades: globalisation. But he does this rather implicitly. This chapter argues that in the same way that strategy had to be developed as a core theme in *Post Normal Accident*, the implications of globalisation need to be further acknowledged and developed than it is done in current studies. In the first section, this chapter comes back briefly on 20 years of controversies over the notion of globalisation, selecting three important ones.

One controversy challenges the assertion that globalisation is a new phenomenon, another question how good (or bad) globalisation is while a last one wonders if globalisation leads to a worldwide homogeneity of societies. In the second section, an integrative framework is developed to connect high-risk systems with trends associated with globalised processes: digitalisation, externalisation, standardisation, financialisation, and self-regulation. The case of BP is again used to illustrate this time the relevance of the framework, and a proposal is made to substitute or complement the idea of network failure accidents to the notion of component failure accidents.

Globalisation: A Very Short Overview

Globalisation, a Central Notion

Globalisation is one of the most important concepts of the beginning of the 21st century. It is at the heart of a renewal of the intellectual background of the social sciences. Retrospectively, it is the collapse of the USSR, symbolised by the fall of the Berlin Wall in 1989, which created both a new situation and an intellectual void to be filled. The primordial questions became: "How does one make sense of the world now that the Cold War is over?" "How does one make sense of the world now that democracy and capitalism had taken central stage over communism?" "What sort of macroscenario could be replacing it in order to help explain the new unfolding world?"

Hannerz (2003) argues that in the competing market of leading theories with the power to structure our understanding of the world, globalisation, in time, prevailed over alternatives such as Fukuyama's "End of History" (1992) and Huntington's "The clash of Civilisations" (1996). Fukuyama argued that liberal democracy was now the ultimate political option on the verge of spreading worldwide. Huntington disputed Fukuyama's argument, insisting that the cultural side of civilisations defines the context in which sociopolitical systems can develop, emphasising the fact that western values would be challenged by other ones and their associated principles of social organisation.

Facing the loss of an encompassing explanatory scheme, the argument is that globalisation provided a substitute to the Cold War narrative. Globalisation, as a concept, was flexible enough to allow for a diversity of interpretations to produce an understanding of a new era (Dufoix, 2013). It could be explored from different disciplinary angles, including economics, geography, anthropology, sociology or political sciences. Writers of globalisation highlight the increasing and accelerating level of interactions in the past three decades, in the 1980s onwards brought about by waves of technological development (e.g. Internet, aviation and maritime transport) coupled with political economy neoliberal orientations (i.e. liberalisation,≈deregulation, privatisation) and increasing flows of people, goods, information, money and images. This, they argue, reconfigures the way societies function worldwide. Let's mention some of these writers.

Stiglitz, an economist, argues, for instance, that globalisation as defined above created the instabilities (e.g. world imbalance and inequalities, deregulated financial capitalism), which led to several economic and financial crises in the 1990s and 2000s (Stiglitz, 2003, 2010); Sassen, a sociologist and economist, contends that globalisation produced the rise of global cities (e.g. New York, London, Tokyo) throughout the world (Sassen, 2006, 2007). Castells, a sociologist, claims that globalised processes unfold through networks that are new configurations of businesses and states as opposed to the more traditional vertical and hierarchical ones (Castells, 2001, 2009), whereas Appadurai, an anthropologist, understands globalisation as a destabilisation of the traditional strong coupling between nation states, cultural identities and geography (Appadurai, 1996, 2006).

A First Controversy: Is Globalisation Really New?

But the notion of globalisation is criticised by the so-called sceptics who argue against those they consider to be globalists (writers indicated earlier). The sceptics argue that globalisation is not new and that the phenomena described by the authors have already been witnessed in the past. In their view, this is not the first time the world has gone through a stage of intensified interactions of flows. One case in point is the period before the First World War, in the 19th century, when maritime transport, communications and economies flourished between western nations and beyond, through political, scientific, industrial revolutions and European empires (see, for instance, Berger, 2003).[1]

Sceptics also argue that globalisation is very asymmetrical, beneficial only to some at the expense of others, connecting some parts of the world (e.g. the triad North America, Western Europe and East-Asia) while leaving a substantial part outside. The argument is that the notion of globalisation poses a problem of definition if it only concerns a restricted proportion of the world population. They can for this purpose refer to Wallerstein's world systems view. He has argued since the 1970s and 1980s (before the globalisation debate) that the world is primarily capitalist, in which one finds core, semi-peripheral and peripheral economies (for an overview, see Wallerstein, 2004).

On the opposite, leading theorists of globalisation are precisely those who see something else. They see instead a quantitative and qualitative change from the 1980s and 1990s onwards through not only the development of information and communication technology (ICT) and transport, as well as the Reagan–Thatcher (neoliberal) politics of privatisation, liberalisation of trade,

[1] And an even broader picture of globalisation seen through a historical lens would go back in time to very different key moments in the history of humanity (Therborn, 2011).

market and finance, but also deregulation that triggered these globalised flows. For these authors, there has been in this respect a notable transformation in the way the world functions observed in the last decades of the 20th century and the first decades of the 21st century. They consider globalisation to be a very meaningful concept for our understanding of the current situation, with its profound transformations in communication exchanges, cultural-national identities, migration flows, economic dynamics, political institutions and global governance.

A Second Controversy: Is Globalisation Good or Bad?

But one fundamental problem is the difficulty to disentangle the normative discourse and the descriptive one. For many, the free circulation of flows across countries underpinning many descriptions of globalisation is an ideologically charged concept translating the ideal of neoliberal markets and societies. An additional and associated controversy is thus whether globalisation is a good thing or not. On the one hand, there are writers seeing the positive outcomes of globalisation and its sustained potential to deliver positive ones in the future (e.g. Goldin & Kurana, 2016; Norberg, 2018). They describe worldwide lifestyle improvements spreading through education, science, technology, decrease of poverty and violence, liberal philosophy and free markets.

On the other hand, there are authors who emphasise globalisation's dark sides (e.g. Sassen, 2014; Guénolé, 2016; Geiselberger, 2017). They see looming ecological disaster, inequality, deregulated financial capitalism and the rise of populisms. For Rodrik, markets, states and democracy cannot harmoniously interact within globalisation (Rodrik, 2011), whereas for Milanovic, the question is an unsettled one, it is impossible to unambiguously conclude that globalisation is good or bad (Milanovic, 2019). Moreover, globalisation which currently unfolds will probably be quite different than it was in the last decade of the 20th century and the first decade of the 21st century, because emerging countries, the BRICS (Brazil, Russia, India or China), each in their own way, create the contours of another globalisation (Ruet, 2016).

A Third Controversy: Is Globalisation a Process of Uniformity?

A last controversy discussed in this introduction is whether or not globalisation leads to a uniformity of capitalisms, businesses and cultural practices under the influence of powerful hegemonic states (such as the United States). Have the drivers of globalisation identified above, namely, deregulation, privatisation and the ICT revolution (creating the increase of flows of people, money, images, ideas, goods across the world) generated a process of homogeneity?

Is there a process of Americanisation, or more broadly of Westernisation of the world throughout the political, economic, social, cultural and technological spectrum? The empirical responses whether from the point of view of the capitalisms (Morgan & Whitley, 2014), businesses (Berger, 2005) or cultural practices (Appadurai, 2006) are unanimous. Despite trends shared through globalised processes by a vast majority of nations (more about this later), diversity is still present.

Despite, from some, the ambition with the ideology of globalisation to create a world free market, countries' capitalisms still maintain their specificities; despite the business trends of outsourcing and creating global value chains, companies differ in this respect according to countries and products or services; despite the cultural influence of soft powers such as the United States, hybridity and resistances are the norms across the world. And, what was already witnessed by the end of the 1990s and early 2000s when the United States and western world was dominating is now even more relevant with the rising status of new powers (Zakaria, 2012).

Beyond the Controversies

It remains that whether one is sceptic or globalist, whether one has a positive or negative view of globalisation or whether one supports or expects a different globalisation in the future through the rise of new powerful nations and geographic areas, one issue remains to translate it more explicitly than is available today into the study of high-risk systems.

One problem is that the contributions of the leading authors of globalisation studies as mentioned earlier cannot be directly applied to this topic because they have not been designed for this particular question, but a translation is possible and needed. The next section introduces one option, proceeding with a combination of studies available but scattered to show that these are in fact connected rather than independent. But let's illustrate first how other research areas have translated globalisation.

Connecting Globalisation to High-Risk Systems

Other research fields have indeed already integrated these changes, which offers insights about how to proceed. Thus, in studies of organisation and regulation, one finds expressions produced in the past 15–20 years such as "post-Fordism" (or post-Taylorism & Veltz, 2008, or postbureaucracy, Alvesson & Thompson, 2006) and the "postregulatory state" (Scott, 2004), all of which translate empirically in their area of concern the implications of this new historical moment and conditions for different kind of actors and topics. This corresponds to middle-range research strategies. They do not conceptualise globalisation directly but investigate its implications for existing and established research topics (e.g. organisation, regulation). Let's present them briefly.

Postbureaucracy

Many for-profit organisations and businesses have found themselves in globalised contexts in the past two to three decades. These include extended financial environments and greater exposure, worldwide competition, work and labour flexibility, incentives to breakdown vertical structures to gain flexibility through novel and expanding ICT networked infrastructure, normalised practices and dependence on a growing service activity (e.g. consulting). This is what the expressions post-Fordism, postbureaucracy and post-Taylorism refer to. However, it remains to be seen precisely, in specific cases, how this is concretely experienced. Difference is always to be expected considering the range of engineered systems, businesses, and social and national contexts, despite the sharing of globalised transformations (e.g. Berger, 2005; Morgan & Whitley, 2014).

Postregulatory State

Similarly, public policy and state regulations have been moving from being centralised, with national laws and authorities dominating the scene, to a condition described as governance in which private (e.g. normalisation bodies such as the International Organization for Standardization [ISO], consulting firms) and nonprivate (e.g. nongovernmental organisations [NGOs], professional associations and also, for European nations, EU laws) actors enter the game, in which practices are framed not only by laws made by nations but also by standards in self-regulatory schemes. This is what the concept of "postregulatory state" (Scott, 2004) (or "governance without government" [Rosenau, 2003; Vogel, 2008]) refers to. But here again, the extent to which and how precisely this configuration applies for specific regulatory domains and across nation states requires empirical descriptions (e.g. Drahos & Braithwaite, 2001; Hay, 2006).

Studies on Globalisation in the Field of Safety

There is no equivalent in writings concerned with the safety of high-risk systems. Studies in this area are a few decades old, but none have addressed this phenomenon broadly so far. This does not mean, however, that there is nothing about the important transformations of the past three decades that is available in the literature. In fact, it is possible to find studies dealing with some aspects of globalisation and its implications. Empirical and conceptual works exist but remain unconnected. A bigger picture linking them to globalisation is not yet established but is needed. One strategy to overcome the situation is nevertheless possible as Figure 4.1 visualises it.

At the top of the picture, globalisation is divided along the three main drivers of the 1980s and 1990s, and then throughout the 2000s identified in most prominent authors' writings very briefly introduced earlier:

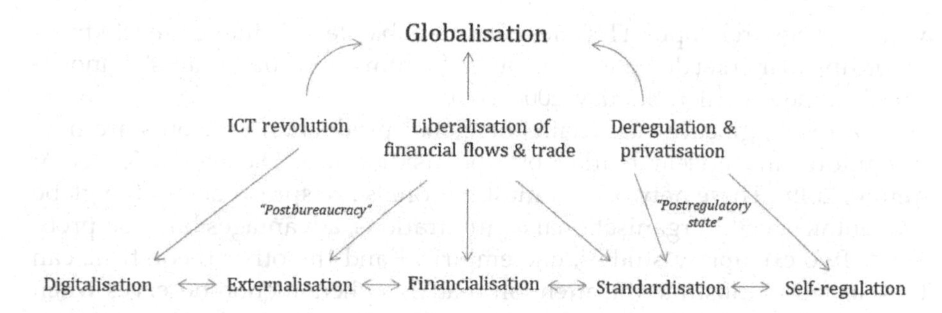

FIGURE 4.1
Globalisation and trends shaping a new landscape of high-risk systems.

ICT (and transport) revolution; liberalisation of finance and trade; privatisation and deregulation. These three main drivers of globalisation are then again divided in five trends that are interacting, both products/producers of globalisation and present today in high-risk systems: digitalisation, externalisation, financialisation, standardisation and self-regulation (their order of presentation does not matter; they are intertwined, amplifying and parallel trends).

This simple structure, which is also an integrative framework, affords the possibility to now connect many dispersed studies that explore only one aspect of the mutations of the operating landscape of high-risk systems and their implications in relation to globalisation. This structure remains compatible with the notion of postbureaucracy and postregulatory state developed in organisation and regulation research traditions, as indicated in Figure 4.1. Let's now comment successively several studies, which fit in these trends, and then show how they are intertwined and not unconnected to each other in high-risk systems.

Externalisation

Externalisation (also called outsourcing) is one core trend associated with globalisation, a product of both a new managerial mindset and fashion that has embraced the idea that organisational flexibility depends on modular or networked principles in a globalised market which made it possible (Kunda & Ailon-Souday, 2006). It is allowed through opportunities offered by transport, the ICT revolution and new flows of foreign investments combined with labour costs.

Many companies have been redesigned according to these principles advocated by managers and consultants. "They envision the organization as consisting of flexible collections of small, self-contained yet interrelated units, free-standing 'businesses', or horizontal, multifunctional, at times even interorganisational teams. Organised around key processes or shifting projects, these self-contained units, 'businesses' and teams should, in this view, strongly rely upon IT systems that enable flexible interconnectedness according to market demands, customers' whims, and competitors' manoeuvres" (Kunda & Ailon-Souday, 2006, 204).

As a result, global value chains or global production networks are now ubiquitous and a clear marker of globalisation (e.g. Dicken, 2015; Coe & Yeung, 2019). These networks of multinationals are spread across the globe and entail specific organisational configurations, advantages but also problems. Two exemplary studies, one empirical and the other theoretical, can be selected to illustrate the attention that this phenomenon deserves when it comes to its implications for safety. On the empirical side, Quinlan, Hampson and Gregson (2013) have established that based on the investigation reports of the National Transportation Safety Bureau, the outsourcing of aircraft maintenance is one of the causes of serious accidents and crashes in the United States.

Behind many events in the past 10 years, problems of maintenance can be directly linked to subcontracting strategies and realities. Their article is also a critical appraisal of the regulator, the Federal Aviation Authority, for not adapting quicker to the changing risk profile of the industry. Despite warnings about the established links between these events and new organisational configurations, regulators were slow to act.

From a more theoretical angle, Milch and Laumann (2016) compile and exploit their findings from a literature review on the link between outsourcing and safety. They found 22 articles on this topic that they discuss in terms of common themes across the literature, retaining "economic pressures", "disorganisation", "dilution of competence" and "organisational differences" as important issues. They conclude, consistent with Quinlan and his coauthors' analysis, that "such issues may contribute to the occurrence of organisational accidents" (Milch & Laumann, 2016, 16).

But outsourcing which characterises the networked properties of high-risk systems must be understood from a wide range of externalised activities, including consulting (Clark & Kipping, 2012). Companies rely indeed on experts in a diversity of domains, from legal, financial, logistics, managerial to safety ones. The contributions of these outsourced services to the operations of safety critical organisations are a core aspect of globalisation, also translated in standardisation to which we turn next.

Standardisation (and Bureaucratisation)

The second trend of the first approach to globalisation and safety research (which consists of identifying a specific topic but not explicitly connecting it to globalised processes) is standardisation (Ponte & Gibbon, 2005). Standardisation has been expanding in the past two to three decades. Busch considers standardisation to be a very important aspect of our contemporary world in general. Social spheres as diverse as science, medicine, education, families, fashions, factories, communications, management and the military are now standardised; in other words, "every conceivable aspect of life now has standards associated with it" (Busch, 2011, 195).

Although dating back to the Enlightenment, Busch also insists on the role of standards in the development of globalisation "the very standardisation of the market—its goods and its institutions—was directly linked to the increase in world trade" (Busch, 2011, 112). This is what Brunsson and Jacobson also note "people and organisations all over the world follow the same standards. Standards facilitate co-ordination and co-operation on a global scale. They create similarity and homogeneity even among people and organisations far apart from one another (...) it can be seen as involving too much regulation, for example, or too little, of not being sufficiently democratic or of leading to undesirable rules" (Brunsson & Jacobson, 2002, 1, 9).

Because we rely on a world composed of external parties bringing a diversity of products and services from around the world, standards have become a core process by which trust is secured. Indeed, standardisation goes hand in hand for this reason with certification. It is the process that ensures that standards are complied with and is "a form of advice to others" (Busch, 2011, 211). Relying on external auditors to verify compliance to standards, "proponents of certification argue that it eliminates surprises by ensuring that people and things act in the manner desired by the buyer" (Busch, 2011, 211).

Two articles can be selected, one empirical, addressing the problem of standardisation (combined with certification) in relation to professionals' practical knowledge (Almklov, Rosness, & Storkersen, 2014), and the second, more theoretical, which includes standardisation (and certification) in a discussion about the phenomenon of the bureaucratisation of safety (Dekker, 2014). Almklov, Rosness and Størkersen compile the findings of two studies they have conducted in two safety-critical domains, maritime industry and railways, to assert that recent organisational trends have undermined local practices of experts in both industries.

By rationalising and normalising safety management systems with the help of consultants, they have decoupled these systems from daily life and challenged the possibility of voicing concerns about safety as was possible in the past. They have structured new power relationship between actors. Evolution in regulation and an increased reliance on consulting firms combine to shape this evolution: third-party auditors and certification bodies play a considerable role in formulating expectations in terms of safety practices, something that was explicitly criticised very early by Power in his thesis of the audit society (Power, 1997).

Dekker defends a similar thesis, building the case more theoretically through a literature review of what he calls the bureaucratisation of safety (Dekker, 2014). He identifies regulation, insurance, contracting and technology as drivers of this current state of affairs. His classification of potential negatives and drawbacks associated with bureaucratisation includes the inability to predict unwanted events, "number games", structural secrecy, pettiness and the hampering of innovation. This idea is further expanded by this author through an anarchist critic of what is framed as an "authoritarian high modernism" (based on Scott, 1998) and applied to the domain of safety (Dekker, 2018), following Wears and Hunte (2014). But these examples do not link the themes they explore to wider trends of globalisation, although it would be relevant to do so.

Financialisation

The third trend described is financialisation (Epstein, 2009; van der Zwan, 2014; Kay, 2015). Financialisation is the new context to which capitalism, to various extents across the world, has been exposed in the past three decades. Shareholder value, dominance of capital market over bank-based system, explosion of financial instruments for trading derivative securities and profits through market rather than commodity products have promoted a financial mindset and influence (Epstein, 2009).

Financialisation is the process by which thinking through such a financial mindset penetrates deeply and structures a various range of areas from banking to industry and households through states (van der Zwan, 2014). This situation has been made possible through several innovations of the late 20th century such as breakdown of the financial architecture, the creation of new market in derivative securities, the development of the mathematics of financial markets, regulation and deregulation, the support of free market and information technology (Kay, 2015, 17).

Implications are great because it reshapes decision-making processes in many spheres and expose banks, households, industry and states to greater risk of financial excesses in the absence of adequate regulations, as witnessed in the 2008 crisis (Stiglitz, 2010). Reshaping corporate governance by increasing the influence of financial actors (e.g. hedge funds) in companies' managerial decision-making processes is one central consequence of this for businesses, including incentives given to CEOs to increase shareholders' return on investment (ROI) favouring a logic of short-term gains rather than long-term ones (Murphy & Ackroyd, 2013; Dumez, 2014).

Another aspect of the financialisation of the global economy and capitalism (although it is not homogeneous across continents and countries) not only tends to be the mergers and acquisitions (M&A) regularly attempted by companies to reach a size that shelters them from aggressive takeovers by bigger than them on the stock markets but also to please investors by higher ROI by mutualising activities (Gomez, 2013). Other strategic options triggered by financialisation of businesses are externalisation, reengineering and downsizing, all of them providing higher ROI to shareholders. Multinationals' business and market contours depend therefore on a complex mix of industrial strategies coupled with financial ones in order to satisfy shareholders.

The implication in the case of high-risk industries is obvious but rarely addressed in empirical research. If short-term perspective is valued more than the long-term one, the risk is to refrain investing in the costly inspection and maintenance of hazardous installations. The risk is to favour cost cutting for short-term return at the expense of investing in safety, which requires longer-term perspectives. An exception is found in Maslen and Hopkins' article and book about safety incentives and rewards although again without explicitly referring to the new operating landscape that it is a manifestation of. Investigating how managers were assessed, they conclude that "financial and business performance held the greatest weighting, representing 70–90% of the total" and specify that "financial incentives can drive decisions that are not in the long term interest of an asset and/or company were common" (Maslen & Hopkins, 2014, 422).

Digitalisation

The fourth trend discussed is digitalisation, a multifaceted transformation of the social world that connects the local with the global in unprecedented ways and at the heart of profound transformation of societies including work, organisation and businesses (Cardon, 2019). The first aspect is that it connects employees in companies from different countries in ways unimagined before, an issue associated with the already-mentioned topic of multiple organisations (externalisation). These connections are strongly mediatised by an increasingly pervasive digitalisation of work practices, and as a result "members come from different cultures, ethnicities and communities; social, educational and technical backgrounds; as well as from a myriad of companies and countries, making the establishment of highly reliable symmetricity, reciprocity, and muliplexity difficult, at best" (Grabowsky & Roberts, 2016, 4).

Another specific trait of digitalisation is a progressive colonisation, extension, combination or sometimes replacement of human activities by a blend of algorithmic systems and machines (or robots), which can perform an increasingly wide range of manual and cognitive tasks more and more independently, or in cooperation with humans (Frischmann & Selinger, 2018). This colonisation by machines is not limited to some operators' jobs because a very wide range of technology is available to support managerial tasks as well (for instance, financialisation as we know it today has been made possible by trading algorithms replacing portions of traders' activities). This creates the presence of information infrastructures extensively coupled with human activity throughout organisations. But these new configurations are not without generating safety concerns.

For instance, extracting petrol in harsh environments without manned platforms is a technological possibility and prowess that puts into question the degree of control that humans can maintain over remotely performed activities. This is called integrated operations. For Haavik, "information infrastructure is in constant risk of becoming a detached artefact of limited relevance, rather than integral to the work" (Haavik, 2016, 7). One problem is that this decoupling is potentially error prone when not enough anticipated in design, as illustrated in recent accident with autonomous vehicles' experiences by Tesla or Google (Banks, Plant, & Stanton, 2018), and also in the case of software implementation in the new Boeing 737 Max. This issue is also linked to regulatory issues, to which we turn.

Self-Regulation

As a result of the process of deregulation and privatisation coupled with the liberalisation of trades and finance, the principle of self-regulation has gained strong support. It is the ideal, in a neoliberal ideology and global production networks, of industries that regulate themselves through the production of their own standards, an ideal that goes hand in hand with a "less state" horizon for multinationals operating across the world through many states jurisdictions.

But self-regulation has also an origin in a regulatory alternative to the command and control philosophy. The command and control regulatory option can indeed generate many problems when detailed legal prescriptions are, for instance, more difficult to establish. Because of fast pace of technological innovation, it can be difficult for states to keep up with it. When prescriptive (rules to follow) or outcome-based regulations (specified targets to reach) are impossible, process or management-based regulations are implemented. They require companies to justify through systematic programs that they put in place proper internal controls (Parker, 2002).

But self-regulation is never totally achieved, and a diversity of versions exist along a continuum between external supervision by states and internal control by companies, leading to a complex mixture of laws and standards controlled by a blend of states, private firms and nonprofit associations. One result is, however, that various actors contribute to the regulations of high-risk systems, from states to for-profit organisations through standards promotion (such as the ISO) but also NGOs.

Regulations have for this reason become quite complex and networked (Mills & Koliba, 2015), and many aspects are worth addressing when it comes to safety when considering self-regulation. Johnson points to some of them. First a philosophy of self-regulation can, in time, lead to "an erosion of the 'underpinning technical knowledge' across many agencies" (Johnson, 2014, 158), and second, it can be associated with "a shift in regulatory focus from the detailed technical analysis of safety-critical systems to a more superficial focus on auditing and assessment of compliance with standards that in many cases have not themselves been subject to detailed validation (…)" (Johnson, 2014, 157).

This last point participates in the trend of standardisation and certification, by which some options of standardising are widely applied from the technical to the managerial spheres. By standardising management (e.g. safety management systems), regulatory activities (or certifiers) rely on management-based philosophies consisting of controlling mainly paperwork and simplified issues of organisational life (a subject discussed extensively previously by Power, 1997, 2007) and are connected to the drawbacks discussed earlier through standardisation.

Combining Studies

If some of these authors indicate globalisation trends, for example, Almklov, Rosness and Storkersen who write that "deregulation and international competition is a key driver toward this trend of standardisation" (Almklov, Rosness, & Storkersen, 2014, 31), they do not explicitly refer to its key features as described in the literature on globalisation, but only address one of these trends. One can establish that topics such as "standardisation" and "externalisation" studied separately and independently by authors of the first approach are, however, strongly linked, something hinted at by Dekker (2014, 2018), but not fully developed.

One can certainly state that standardisation is a product of globalisation, as do Brunsson and Jacobson: "globalisation creates an increase in the demand of world-wide standards" (Brunsson and Jacobson, 2002, 9) or Busch "the very standardisation of the market-its goods and its institutions-was directly linked to the increase in world trade" (Busch, 2011, 112). Extended commodities' chains, relying on a diversity of separate entities linked through global networks of people, goods, information and money, have resulted in standardisation as a solution to ensure that the expectations of various contracting organisations are met, to ensure trust.

By certifying that organisations have met standards, they can enter globalised markets while providing the assurance that they will deliver products with the required level of quality, safety, etc. On top of this, deregulation of some activities (a salient feature of the neoliberal ideology supporting the discourse of globalisation) has favoured self-regulation regimes, which also rely on the standardisation of management-based principles to comply with authorities' expectations (Parker, 2002).

This situation has been coined a "neo-liberal paradox" (Hibou, 2012). The paradox is expressed as follows: although flexibility is advocated in a highly competitive market (the so-called 'postbureaucracy' principles), standardisation has gained ground to the extent that it sometimes burdens organisations, impeding the very flexibility initially advocated (Dupuy, 2011).

One could add that financialisation is translated into accounting numbers serving the appreciation of companies, which relies on the power of digitalisation and turns the concrete work of individuals into simplistic, sometimes even invisible realities with the help of consulting firms (Gomez, 2013), and which now leads issues associated with algorithmic regulations (Yeung, 2017), with implications with regard to standardisation (Almklov & Antonsen, 2019). So these trends are not isolated, and they are strongly connected instead and should be considered together not in isolation, as represented in Figure 4.1.

Back to *Normal Accident*

A New, *Post Normal Accident*, Operating Landscape

Taken together, postbureaucracy and postregulatory state, their associated trends as studied in safety research, digitalisation, externalisation, standardisation, financialisation and self-regulation translate concretely the globalised operating landscape, which has unfolded progressively over the past two decades to three decades, following the publication of *NA*. In the 1980s then onwards, as explained in Chapter 1, Perrow relies on the decomposition of technology (task), structure, goal and environment of organisations, a decomposition also at the heart of Hopkins' narrative of disasters, and of his normative model of safety, as explained in Chapter 2.

And, they both agree that some configurations of these four dimensions are more likely to maintain safe practices than other. Such an assessment cannot these days be decoupled from the trends introduced earlier. They have been shaping high-risk systems performances, but did not exist at the time of *NA*, and have therefore not been much reflected upon. Let's connect the trends introduced earlier briefly in relation to Perrow's and Hopkins' analyses of organisation.

First, technology (and task) has been reconfigured by digitalisation and standardisation. There is no area in which computers based on increasingly algorithmic and artificial intelligence cannot potentially play a role to assist and sometimes even replace people's tasks. Although always a matter of choice, it seems now that automation as encountered with computers after Second World War was only a step towards a greater level of cognitive assistance and cooperation or perhaps replacement by machines through digitalisation. But digitalisation also combines with standardisation of work to create new configurations of tasks for people, and is not independent trends.

Second, structures of organisations can now exhibit strong networked properties relying on externalisation along principles of flexibility and autonomous business units (BUs) across continents because of the possibilities offered to them in this respect through technology and more open borders. Although this again appears to be a matter of strategic choices for companies (as automation and digitalisation is), the network configuration is now ubiquitous and largely applied throughout industries.

Third, goals of companies, their strategies, must be formulated in financialised and potentially increasingly self-regulated type of environments, which are, fourth, themselves now increasingly global markets. The fact that companies evolve in these new contexts is obviously not without consequence for their strategic decision-making processes and their response, as global production networks, to opened markets fraught with opportunities as much as risks (Yeung & Coe, 2014).

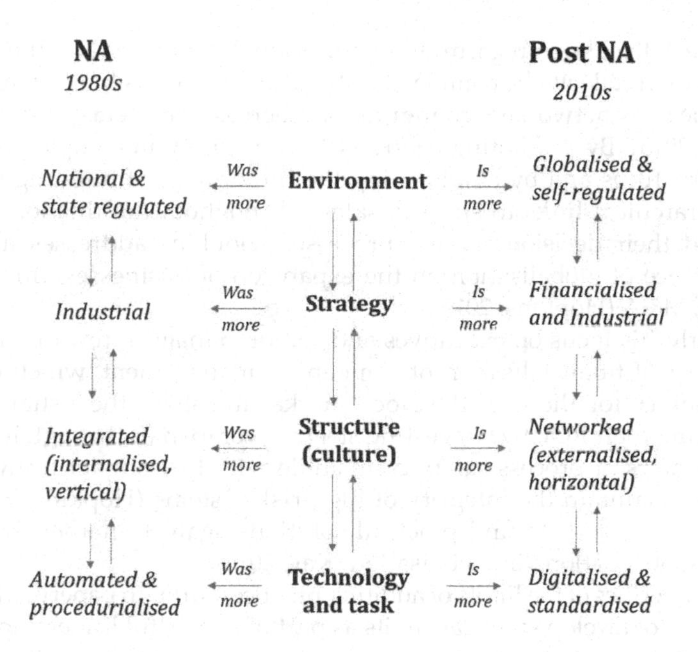

FIGURE 4.2
From *NA* to *Post NA*, a new operating landscape. NA, Normal Accidents; Post NA, Post Normal Accident.

With a bit of simplification to highlight the message, this new situation contrasts with the background of empirical and conceptual studies available in the 1960s, 1970s and 1980s, which influenced Perrow when writing *NA*. Technology and tasks were more automated and proceduralised than digitalised and standardised, structures were more integrated than networked, strategies were more industrial than financialised and environments of organisations were more national and state regulated than globalised and self-regulated. This oppositions meant to grasp these multidimensional changes can be illustrated as follows (Figure 4.2).

Hopkins' View and Globalisation

So if the decomposition between technology (and tasks), structure, goal and environment still provides an adequate and useful set of analytical principles to study high-risk systems, much has changed since the publication of *NA*, but not much reflected in Perrow's subsequent work (apart from Perrow, 2015 on self and deregulation). These trends are more visible in Hopkins' studies although in an indirect manner. His analyses and recommendations target indeed many of the processes that have structured companies under globalised trends over the past 30 years. Let's comment this briefly.

Hopkins' discussion regarding organisational structures and the need for centralisation reflects the complexity of multinationals as they evolved from hierarchical to networked configurations across continents and branches (Dicken, 2015). By promoting centralisation of safety in complex organisational structures and by emphasising the importance of building matrixes with hierarchical links to support safety throughout multinational operations and their decision-making processes, Hopkins addresses indirectly the influence of globalisation on the expansion of businesses, through, for instance, M&A (Hopkins, 2019).

Similarly, his focus on executives and senior managers' rewards indicates the context of financialisation of companies' management, which depends on the ability for those in the stock market to satisfy their shareholders, maximising their ROI. By suggesting steering companies through incentives and indicators of process safety compatible with the long-term investment needed to maintain the integrity of high-risk systems (Hopkins & Maslen, 2015), Hopkins tries to find practical solutions against another driver and trend of globalisation, financialisation (Kay, 2015).

Hopkins' critics of the limits of auditing practices through paperwork and his suggestion to develop sceptical audits as part of a mindful leadership are also the products of the business of standardisation, certification and self-regulation (Hopkins, 2007a). Standardisation of practices through safety management systems is a response to both the need to establish trust among companies as much as a response to the shift from prescriptive to self-regulated legal philosophies of health and safety in a globalised context (Busch, 2011).

Finally, Hopkins' recommendation to design regulatory regimes based on safety case, namely, a legal requirement for companies to perform their own assessment is a response to this move from a prescriptive to a self-regulated philosophy. Indeed, the absence of prescriptions formulated at the level of the law should not equate with the absence of prescriptive rules in companies. Companies must still devise prescriptive principles that derive from risk assessments defining the engineering and procedural standards to be complied with for safe operations.

These practical responses (i.e. adequate incentives and indicators, centralised matrix structures, sceptical audits, safety case regimes), however, are an aspect of Hopkins' work that one can only fully understand when sensitised to the evolutions as described earlier (Figures 4.1 and 4.2). Hopkins was not unaware of this; he writes, for instance, "many companies in this era of globalisation have restructured into horizontally-aligned independent units and this has often resulted in a drastic cutback in safety management from the center" (Hopkins, 2000, 37). But his practical mindset led him to stress prescriptions rather than broad descriptions, which would make explicit the connection of his work with the implications of globalisation.

Every day, as illustrated with the selected studies earlier, companies operate under these drivers and trends of globalisation, and it is a research program in itself not only to study the effects of these processes on safety in daily operations but also and mostly to understand their combined effects. As Pidgeon asserts it, "it seems unlikely that industrial safety practice can remain completely insulated from these major global developments. This in turn sets a research agenda to fully understand the ways in which safety is indeed being impacted, and an immediate challenge to respond with new approaches and methods of risk governance" (Pidgeon, 2019, 266). As usual, it is much easier to relate to these combined effects in retrospect, when links can be clearly established, as for instance, with the story of BP. This case study is now again used to show this time how it is, beyond its specific history, an example of fiasco of a globalised corporation.

The Story of BP, the Strategic Fiasco of a Multinational

Let's come back on what has already been introduced in Chapter 3 about BP and then connect it further to this chapter on globalisation. Because of the intensive coverage through documented investigations and books, BP's troubles offer an opportunity to illustrate what happened to the multinational during the 2000s and to link it with the trends discussed here. Between 2005 and 2010, the multinational BP experienced a series of highly commented events in different branches of its operations.

The first in 2005 is an explosion in the Texas City refinery, which caused the death of 15 subcontractors on site. The second in 2006, in Alaska, is a release of oil into the environment, which was due to the corrosion of a transport pipeline. The third in 2010 is the explosion of an offshore platform commissioned by BP for the Macondo well, which engendered the biggest oil spill in US history, and, again, the loss of 11 lives, primarily employees of the contracted companies.

One interpretation is bad luck. Because majors in the oil business manage complex sociotechnological operations, one central issue for multinationals is to manage their geographic and product diversity, and accidents are bound to happen from time to time considering the difficulties involved in managing such companies (Morgan, 2006). Three in a short period of time would just be an example of the adage "bad luck comes in threes".

The second interpretation followed in Chapter 3 is that the corporation created the conditions for an increased likelihood of incidents and disasters and that it can be characterised as a strategic fiasco. In a rather short space of time (5 years), three major accidents occurred in the corporation's core activities: refining, oil transport (pipelines) and offshore exploration. To explore to another level this story in relation to globalisation, one needs to summarise mainstream analyses of the case by different authors who converge to describe a company pushed beyond its limits (Bergin, 2012; Lustgarten, 2012; Hopkins, 2008, 2012).

Browne's Legacy

The background of these events was major strategic changes brought about by John Browne, the CEO of the company between 1995 and 2007. BP had become, in the 1980s, a relatively small corporation among giant petroleum companies, which dominated the world oil production (Texaco, Royal Dutch Shell, Texaco, Chevron, etc.), but Browne turned the situation around through a series of moves combining an array of organisational and strategic changes.

As the head of a multinational companies, Browne, throughout the years first secured access to new fields thanks to the development of internal expertise in exploration of deep water wells and by dealing with the social, economic and political contexts in the countries where exploration takes place – in sometimes very complex circumstances including Latin America, Africa and Russia (as discussed in detail in his biography, Browne, 2010).

Second, he bought and merged with other petroleum companies (e.g. Amoco) to grow in size and prevent the risk of aggressive takeovers. Third, he restructured BP's organisation into a networked configuration (the internal engineering expertise was decentralised, BUs were created with a higher degree of autonomy across the branches starting from exploration then to refining), thus increasing decision-making speed and flexibility at the regional/branch levels, and relying on outsourcing (most victims of the disasters are not from BP but from contracted organisations, e.g. Jacobs engineering in Texas City; Transocean, M-I SWACO on Deepwater Horizon).

Along with these strategic evolutions was associated, fourth, a strong cost-cutting policy, which sent signals to markets and investors about the company's ability to increase productivity while at the same time growing in assets and size (in a period of low crude oil prices). These strategies of "acquisition, divestments, restructuring measures, cost cutting initiatives" (Grant, 2003, 510) were actually widely shared at the time by petroleum companies acting under the same globalised trends.

Browne became during the late 1990s and early 2000s a star CEO and voted several times UK manager of the year. His talk on sustainable development[2] made him a "Sun King" (see *Financial Times'* article, Buck & Buchan, 2002) celebrated by the business media, the financial world and even some environmental NGOs. Capturing the ecological momentum of the times, for all intents and purposes, he appeared to be a CEO personifying global consciousness, purporting among multinationals that they had a duty toward a new "green" capitalism.

It is important to underline that, although very successful for a while, Browne's strategy did not appear, overall, to be radically different from his competitors if one follows the analysis of Grant (2003). What was different was the intensity of some aspects associated with this strategy in the case

[2] British Petroleum then became Beyond Petroleum.

of BP (e.g. cost cutting, mergers, networked organisation), combined with a deregulation ideology, especially in the United States.

> Key strategic adjustments by the oil majors included: rationalization of downstream and chemical businesses in the face of chronic capacity (...), refocusing upon core businesses, upstream expansion into new geographical areas (especially China, the Former Soviet Union and Latin America), the adoption of new technologies (e.g., deep-water exploration, directional drilling, 3D seismic analysis, and environmentally friendlier fuels), adaptation to social and political pressures, and responsiveness to the demand of owners.
>
> *Grant (2003, 513)*

When Browne resigned in 2007, partly as a consequence of the conclusions of reports such as that of the CSB on the Texas City event in 2005, a very close collaborator, nurtured under Browne's supervision, Tony Hayward, took over as CEO. He made some changes, including internalising back safety audits expertise, but retained the main strategic features of cost cutting and commercial imperatives under financial markets pressures.

This resulted in a similar emphasis on production over safety in the company (although with different intensity in various BUs and sites, one of the key issues here as the exploration branch did not incorporate the structural changes of BP [Hopkins, 2012, 101]). Hayward was aware that cultural change was needed but did not manage to lead this change to term into the corporation as a whole.

This said, although it occurred under Hayward's watch, the Deepwater Horizon blowout had roots in Browne's strategic program and imprint. These changes with consequences for safety, especially in the United States where BP had grown tremendously in size after several takeovers, including that of Amoco, did not go unnoticed by legal institutions in charge of regulating environmental activities, such as the Environmental Protection Agency. Lustgarten's text is quite telling about the problems the multinational's employees (often subcontractors) faced when trying to intervene in what appeared to be inappropriate safety practices in Alaska in inspection program of pipeline corrosion.

Indeed, many incidents reported by ex-employees and whistle-blowers indicated important problems, which were finally settled through a number of legal arrangements over the years between the multinational's lawyers and the State, in a context of deregulation or "less state" ideology in the United States as hinted earlier. It is interesting to note that while an organisation is active in many geographic areas in the world, the three disasters occurred in the United States.[3] One possibility is that it is related to the company's high number of assets there, but another is that it is also related to deregulation in the United States (e.g. Coglienese, 2012).

---

3 The Grangemouth refinery in the UK also suffered a number of incidents in 2002 which were investigated at the time by the Health and Safety Executive (HSE).

The Networked Firm

It is obvious now, with the background available, how Browne's strategy can be associated with these transformations, including ideologies, management fashions and mindsets applied to organisation and regulation which affected businesses and states in the late 1980s onwards. Using the wealth of data produced and analyses available on this case, one realises the extent with which a company such as BP embraced the idea and put in practice an extreme version of the networked corporation.

As an illustration, in the Macondo project in 2010, a very high number of activities were outsourced and contracted to other organisations, a model replicated in other branches of the group exploiting pipelines and refineries. In the case of Macondo, the network of organisations included Transocean (drilling operator), Halliburton (cement provider), Cameron (blowout preventer), DRIL-QUIP (provider of materials such as casing hanger, seal assembly), M-I SWACO (mud and spacer workforce), Shlumberger (cement evaluation), Sperry Drilling or Weatherford (float valves, centralisers).

Using the global production network analytical framework (Dicken, 2015), Bridge writes that "large upstream operations involved more than one equity partner (to spread risk) and comprise a number of specialist firms to whom different work processes are outsourced. (…) Drilling operations are often outsourced to a contract drilling company who may also provide the rig or drill-ship (e.g. Parker Drilling), and who undertakes to crew the rig. Drilling tool supply may be contracted to a specialist toll company (e.g. Baker Hughes), with data logging, data analysis and well-maintenance contracted to another firm (e.g. Schlumberger)" (Bridge, 2008, 400).

Some of these contracted organisations were and are therefore not small companies but very large ones instead, sometimes worldwide leaders in their markets (e.g. Transocean). On the offshore platform, only 7 people were employees of BP out of 126. This surely reflects the industry of oil and gas more widely and has been so for many years, but this diversity of actors shaping daily operations of the multinational in its branches is only a fraction of other categories of actors directly linked to BP's activities at corporate and strategic levels.

In the narrative of Bergin (2012), one reads about the important roles of other type of actors such as investment bankers, traders, financial analysts, consultants (e.g. Mc Kinsey, Bain and Company), universities (e.g. Harvard, Stanford), control authorities, governments in the United States and the United Kingdom, nongovernmental companies (e.g. ONG), medias and professional associations (e.g. American petroleum associations).

These are of course not all outsourced organisations but reveal how environments of companies have evolved in the past decades. Financialisation, standardisation, digitalisation and self-regulation schemes have indeed created a new world of different actors (e.g. consultants, traders, NGOs) who shape the strategic decision-making processes of executives and top managers which shape in return organisations' behaviours.

Thus, for some, by following consultant recipes based on the idea of decentralised organisations applied in the car industry to petroleum industry, Browne, BP's CEO, would have followed a bad advice. "McKinsey again suggested ditching the 'matrix' structure – which the consultancy had itself devised for Shell in the 1950s – in favour of the decentralised business model" (Bergin, 2012, 17). Hopkins adds that "the change was a commercial success, but the seeds were sown for both the Texas City and the Macondo disasters" (Hopkins, 2012, 101).

Such appreciation is quite a common view of consultants who are often targeted for their lack of positive contribution to companies despite their highly valued status (Sennett, 2006; Sturdy, 2011). One now starts understanding in light of this description why BP constitutes an interesting case of a multinational exemplifying the trends that have been discussed in this chapter.

Translated into a number of different terms in organisation and regulation studies such as "postbureaucratic" and "postregulatory state" with trends (Figure 4.1), they provide a set of interconnected and interwoven dimensions that one can now link to the BP case. First the BP story can be understood in this respect as a "postbureaucratic" and "postregulatory state" failure. Second, it can be refined and decomposed further through the trends identified and described.

Let's discuss these two analytical options. As described earlier, BP developed and expanded under Browne's (then Hayward's) strategy, embracing the new trends, framing what would become, from the viewpoint of industrial safety, a networked organisation ("postbureaucracy") pushed to its limits under an ideology of deregulation policies in the United States ("postregulatory state").

A Postbureaucratic Strategy and Postregulatory State Failures

On the one hand, Browne's strategy was to restructure BP's organisation into a networked configuration after merging with several multinationals (the internal engineering expertise was decentralised, BUs were created with a higher degree of autonomy across the branches starting with exploration and then refining), not only increasing speed of decision-making and flexibility at regional/branch levels but also subcontracting and outsourcing.

The resulting lack of centralised expertise created difficulties for the network of autonomous and decentralised BUs to benefit from corporate oversight that would report to top management about safety issues with sufficient clout to challenge, for instance, the cost-cutting objectives in line with the company's strategy. On the other and, while BP was a company active in many geographic areas in the world, the three disasters (Texas City, 2005; Prudoe Bay, 2006' Deepwater Horizon, 2010) occurred in the United States. One possibility is that it is related to the high number of assets there, but another likely possibility is that it is strongly related to a philosophy of deregulation in the United States.

A weak regulator, with insufficient resources, expertise and/or clout to challenge practices of multinationals operating in the Gulf Mexico (offshore platforms) or in different states (refineries), can lead to bad practices unless industry can self-regulate itself appropriately. As Perrow writes "we really have been moving in the United States towards a deregulated state and deregulated capitalism. Indeed, the proportion of economic activity governed by public regulation in the public interest has declined precipitously while private regulation serving private interests has surged" (Perrow, 2015, 204).

Globalised Trends in the BP Case

But we can move one step further and refine the interpretation of the BP case study in relation to the trends discussed to another level of decomposition (Figure 4.1).

Financialisation

Privatised under Thatcher's neoliberal policies in the 1980s, BP then ventured in the mid-1990s into very ambitious and challenging takeovers and mergers (financialisation). These turned BP under Browne' lead, in the early 2000s, into one of the biggest petroleum companies in the world in terms of assets, after absorbing Amoco, among other companies.[4]

Digitalisation

The power of computers allowed BP to become a champion in exploring uncharted geological territories, locating promising wells and digging deeper and deeper to reach these resources. This is one aspect of the trend of digitalisation. Computers transformed the prospects of petrol exploration, especially deep-water wells, the size and precision of platforms, as much as the networked type of architectures that ICT systems made possible.

Externalisation

Browne's strategy of decentralising BUs towards autonomous self-managed entities was driven by targets and outsourcing, which also included decentralising functional supports away from the matrix organisation (e.g. safety expertise). Decentralising safety meant removing a layer of internal control which ensured the implementation of adequate practices and goals through decision-making processes. This the trend of externalisation, namely subcontracting as a result of companies extending their reach worldwide and relying more than ever on service consulting firms to create complex, networked,

[4] The Texas City refinery was an Amoco asset.

coordinated projects involving contracts and specific relationships between distinct legal entities within and outside the corporation.

Financialisation

As for the industry in general "by the late 1990s, pride of place was given to shareholder return and superior profitability (...)" (Grant, 2003, 512). In BP, finance became the new corporate environment that managers had to satisfy, transformed into quantified targets for shareholders' ROI, which could be satisfied by increased profitability through cost cutting and mergers. This corresponds to financialisation through which market appraisal of companies' results and future prospects driven by shareholders expectations on investment returns shaped CEO's strategies and communication – mergers, business media, profitability, financial targets.

Standardisation

By standardising behavioural-based safety approach practices, safety management systems and occupational safety indicators, BP produced an image of control, a façade (Brunsson & Jacobson, 2002;), based on rational frameworks of procedures and standards decoupled from concrete process safety issues (Hopkins, 2011). This is standardisation, i.e. widespread management system principles promoted by consulting firms and certification schemes in global markets, which challenge professional expertise and simplify reality while bringing confidence to managers.

Self-Regulation

Finally, weakened regulatory oversight in the United States raised concerns about the ability of the state to achieve control of complex sociotechnological systems requiring expertise, resources and power, which created a favourable environment for BP to operate contrary to safe practices. This is the result of self-regulation, partly the product of the neoliberal ideology favouring less state and less inquisitive regulation and authorities in favour of a relatively self-regulated industry relying on standards developed within the industry, consulting firms or international normalisation bodies (e.g. ISO).

From Component Failure to Network Failure Accidents

What is particularly interesting with the BP case is the example it provides about what this chapter has discussed. Globalisation has shaped a new operating landscape for high-risk systems. Digitalised, externalised, standardised, financialised and self-regulated following liberalisation of trade and finance, privatisation and deregulation combined with ICT and transport revolutions, high-risk systems are best understood as complex networked high-risk systems.

One can see some key categories of actors representing the trends introduced in this chapter in BP's description such as externalisation (products and services providers), financialisation (traders, financial analysts), self-regulation and standardisation (consultants, regulators, standards organisations, NGO) defining the configurations characterised as networked. However, and as argued about globalisation, in general, in the first section of the chapter, some of these globalised trends are not entirely new, not good or bad, not deterministic forces that create convergence between organisations and depend on strategic choices.

They are nevertheless sufficiently pervasive to constitute a widely shared template which influences technology (and task), structure, goals and environment of organisations as described by Perrow in the 1970s and 1980s, or Hopkins later in the 2000s and 2010s as found in his accounts of accidents. Technology and tasks are potentially increasingly digitalised and standardised, structures of organisation are potentially increasingly externalised and decentralised, more self-regulated and standardised too, and goals and environment of organisation are also potentially and increasingly financialised and globalised (Figure 4.2).

The virtue of a template at this very broad and generic level of description is to sensitise more than impose a rigid interpretive canvas. It is only through empirical ethnographic work that one can explore the concrete implications of these trends in specific case studies while keeping in mind the importance of looking for their combined effects rather than into their isolated manifestations using strategy as a core lens to proceed.

One can see the presence of these trends in many analyses, as in a recent account of pipelines' disaster in the United States (Hayes & Hopkins, 2014) or in daily operations in the maritime industry (Almklov & Lamvik, 2018). But more needs now to be done to incorporate these explicitly in case studies across levels. This canvas does not imply that only bad implications come out of these trends but that it constitutes instead the new operating landscape of high-risk systems.

With this in mind, it seems appropriate, returning to Perrow's terminology, to substitute the idea of component accident with the notion of network accidents. Component accidents were these accidents in Perrow's framework which couldn't be considered as normal accidents produced by tight coupling and interactive complexity. Component accidents were accidents resulting from flaws in the DEPOSE system of companies (design, equipment, procedures, operators, supplies and materials, and environment).

The notion of component failure accidents was the expression chosen by Perrow to describe his second thesis. Considering now that many organisations exhibit networked features captured by ideas such as postbureaucratic and postregulatory systems combining digitalisation, standardisation, externalisation, financialisation and self-regulation, flaws in the DEPOSE system of companies are likely to be best understood as some of the consequences of these networked features now exhibited by high-risk systems.

Summary of Chapter 4

Introduced at the end of the 20th century, following the end of the Cold War, the concept of globalisation provided a new encompassing perspective to think and to link some of the major changes societies have been facing in the past decades. It is a central topic for the social sciences which contributed through a number of controversies to a better understanding of what such a broad notion entails.

Globalisation has transformed the operating landscape of high-risk systems, but it has not been made an explicit dimension of safety and accident research so far, although translated in other domains in notions such as postbureaucracy or postregulatory state. There are nevertheless some studies in the field which address some of the trends associated with the consequences of globalisation on organisation and regulation. They discuss core issues of digitalisation, externalisation, standardisation, financialisation and self-regulation.

When put together into an integrative framework, these interwoven trends help sensitise Perrow's analytical categories of technology (task), structure, goal and environment (also found in Hopkins). Digitalised and standardised technologies and tasks, externalised structures, financialised goals (strategies) and self-regulated environments constitute new networks configurations of companies that an organisation such as BP embraced in the course of the 2000s through its strategy with dramatic consequences.

Shaping the networked properties of such configurations, these new trends which have been studied separately must now be integrated into a broad view of how safety performance is produced and maintained in high-risk systems in this new operating landscape. Substituting the notion of network failure accident to the notion of component failure accident developed by Perrow in *NA*, this chapter adds another dimension to the *Post NA* narrative.

5

(Global) Eco-Socio-Technological Systems: Expanding Scale, Scope and Timeframe

Introduction

Perrow's matrix of (loose/tight) coupling and (simple/complex) interaction is the graphic of *Normal Accident* (*NA*) visualising normal accident (na): some systems are more likely to surprise us than others because of their structural features of tight coupling and complex interactions (Figure 1.1, Chapter 1). It is also the picture putting together and classifying for the first-time different kinds of high-risk systems. This picture probably contributed to lock readers into Perrow's first thesis, at the expense of his second thesis.

This enduring representation of the 1980s was indeed embracing the technological determinism of the first thesis. A position in top right corner of the matrix meant a higher degree of likelihood of disaster. Another kind of visualisation was presented in the previous chapter with the help of Hopkins (AcciMap, Figure 2.1, Chapter 2), supporting this time the second thesis, some 20 years later, in the early 2000s. It exhibits indeed a multidimensional causality connecting different layers of analysis, far beyond the technological rational of the first thesis.

But if this different representation helps move away from the technocentred argument, it also loses the big picture that Perrow's matrix successfully created in his time. How to keep the heuristic visual value of such a matrix while reflecting the contemporary situation? Thirty-five years ago, high-risk systems were a highly sensitive topic; they have not only remained ever since, but they have also been progressively embedded in wider concerns, wider categories of risks. These wider concerns situate them in different scope, scale and time frame, which could not be discussed by Perrow back then.

In this chapter, a new visualisation inspired by Perrow's matrix is proposed to represent some aspects of *Post NA*. To do so, I first distinguish the emergence of several overlapping categories of risks over the past 30 years: socio-technological, systemic and existential risks. I contend that these categories are connected to diverse phenomena: the advent of high-risk and large technical systems in the 1980s, the process of globalisation described in the 1990s and the phenomenon of anthropocene and transhumanism that gained visibility in the 2000s. Together, these categories constitute a substantial modification of how we think of high-risk systems, their scope, scale and time frame.

This modification leads to a revisit of Perrow's 2 × 2 matrix. Scale of governance and magnitude of impact replace coupling and interactions. I further argue, secondly, that complexity as used by Perrow in *NA* must now be reconsidered. When he used this notion in the 1980s, it was mainly linked to a technological argument. Complexity in the new graph of *Post NA* evolves to become a material, ecological, epistemological and philosophical notion, which should now underpin our approach of safety. Fukushima provides an illustration of an eco-socio-technological disaster.

Expanding Scope, Scale and Time Frame

High-Risk Systems and Sociotechnological Risks

Let's start with the category of sociotechnological risks, using the content of previous chapters and adding some new insights. Perrow developed the notion of high-risk systems and technological disasters with *NA*, and the 1980s have been fruitful decades for this topic. Let's come back shortly on this history which was discussed in Chapters 1 and 2 in order to compare it with the emergence of two other risk categories next. Before Perrow's *NA*, in 1978, Turner published *Man-Made Disasters*, the failure of foresight, a book looking into disasters from a sociotechnical perspective (Turner, 1978).

The contribution of Turner at the time was to go beyond an engineering view of disasters and to understand, study and conceptualise these events as an engineering, organisational and cultural phenomena. Accidents are the products of fallible institutionalised views created by a wide range of actors of organisations. In 1979, Three Mile Island created much concern about safety in an already highly sensitised society to the danger of the development sociotechnical systems in general, following several decades of Cold War and its nuclear war threat. Perrow created therefore a momentum as discussed in Chapter 1.

One companion of the notion of high-risk system is the one of large technical system. The concept of large technical systems was developed around the 1970s and 1980s to describe numerous infrastructures (or networks) that appeared to share common features. One pioneer in the field is the historian of technology Hughes who studied electricity networks (Hughes, 1989) and then other type of systems (Hughes, 1997).

This work sparked interest in the research community and gathered other researchers already involved in this area (Mayntz & Hughes, 1988; La Porte, 1991; Gras, 1994, 1997), and concerned "integrated transport systems, telecommunication systems, water supply systems, some energy systems, military defense systems, urban integrated public works" (Joerges, 1988, 24).

In the 1980s, a series of high-profile explosions, crashes and shipwrecks, among which events such as Bhopal (1984), Chernobyl (1986), Challenger (1986), Herald Free of Enterprise (1987), Piper Alpha (1988) and Exxon Valdez (1989), caused the deaths of many people from the companies involved and sometimes beyond. From there, a rich tradition of various research strands has steadily produced a number of key contributions to conceptualise safety and accidents.

One landmark book in this tradition was Vaughan's study of the Challenger launch (Vaughan, 1996). Coming back on the outcomes of the presidential commission, her view of the event is one of a technosocial coupling in which the appreciation of risks is a sociohistorical process shaped by engineers and managers in the context of an innovative technological system embedded in a highly complex, differentiated and networked organisation, the National Aeronautics and Space Administration.

This approach of accidents was developed in parallel over the years by Hopkins who was introduced extensively in Chapter 2. In a series of books, Hopkins produced in-depth explanations from an organisational or a sociotechnical angle, on the basis of the hearings of the official commissions following particularly mediatised disasters (Hopkins, 2000, 2008, 2012), and produced an important practical, complementary, perspective to Perrow, described as an unofficial theory of *NA*.

In the past two decades, Erika (1999), Toulouse (2001), Buncefield (2005), Texas City (2005), Fukushima Daïchi (2011), Deep Water Horizon (2010), Costa Concordia (2012), AF 447 (2009), Sewol (2014) and many other events throughout the world constitute more recent examples of sociotechnological failures in the beginning of the 21st century in a variety of high-risk systems (e.g. nuclear, maritime, offshore, aviation) illustrating the recurrence of major events.

Globalisation and Systemic Risks

I now rely on Chapter 4 to introduce globalisation, while adding new inputs too. The development of high-risk systems and large technical system in post–world war (and the reactions it triggered about the risk of losing control) is part of an intensified dynamic of interactions across continents in the past decades. Liberalisation of trade and finance, privatisation, deregulation, an IT revolution in the early 1980s, reinforced in the aftermath of the collapse of USSR, engendered a completely novel geopolitical, economic and international landscape in the 1990s as explained in Chapter 4.

Characterised by an increase of flows of capital, people, money, images and goods at the world stage, the notion of globalisation captured these changes, a notion that gained prominence in the social sciences debates in the late 1990s and during the 2000s. Many writers and researchers established key concepts to grasp, describe, analyse and interpret this globalised dynamic, with the implications of these transnational mutations (e.g. Sassen, 2007).

Both Beck and Giddens had anticipated this moment of change in societies and combined it with the notion of risk (Beck, 1992, 1996, 1999; Giddens, 1990, 2000). Because nuclear (the Chernobyl's radioactive cloud did not stop at the Ukraine's borders), chemical and genetic then ecological, financial and terrorist risks are cross-border issues, the world must come together to organise itself in order to find preventive solutions. A cosmopolitism is advocated by Beck (Beck, 2002).

Giddens has a slightly different starting point because he approaches the process of globalisation as a result of modernity expanding in every corner of the world (capitalism, industrialism, nation state). But whereas for Beck risks are globalised in the sense that they are not bound geographically, globalisation is for Giddens the source, in itself, of new risks (Giddens, 1990, 2000).

And it is during a peak of intensity of the debates on globalisation that events in the first part of the 2000s, e.g. the Internet bubble (2000), 9/11 (2000), Katrina (2005) or SARS (2003), emerge along with the already-established notion of sociotechnological risks (e.g. Toulouse, 2001; Texas City, 2005). These cases represent four different types of risks: financial, natural, pandemic and terrorist. The OECD report in the early 2000s (OECD, 2003) puts together these different categories under the new terminology of "emerging systemic risks" and represents a landmark document in this respect.

Because of our more than ever interconnected world, any event locally can have ripple effects throughout the world globally (e.g. financial crisis, pandemics, natural catastrophe), but the reverse is also true, because globalisation is a major source of transformation of our ecosystems through climate change, these global changes have localised effects (e.g. hurricanes, rising level of oceans). They are for this reason considered to be systemic.

Systemic risks are phenomena that result from an increase of a diversity of flows, which connects into networks a multitude of nodes whether actors, institutions or infrastructures. In 2006, a report on what is called "global risks", based on a survey of decision-makers and experts established by the World Economic Forum is published (WEF, 2006), and has been every year since. For Arnoldi, "it is no coincidence that a report on global risks subscribes to the concept of systemic risk" (Arnoldi, 2016, 275).

The terrorist attacks in Madrid (2004) and London (2005), the tsunami in East Asia (2004) or Asian Flu (2004) are all part of this report; they constitute other very mediatised events, which, along the subprimes financial crisis of 2008 (Stiglitz, 2010), only reinforce the notion of systemic risks. It is also a moment when the concept of critical infrastructures takes centre stage in some publications, as a reaction to this evolving global context (e.g. Auerswald et al., 2006).

Continuity and sustainability of our contemporary lifestyles depend on the proper functioning of these critical infrastructures and the trust individuals have in their reliability and safety (Giddens, 1990). This topic has become a very important one, partly overlapping, from an empirical and conceptual point of view, with the category of high-risk systems, large technical systems and sociotechnological risk introduced in the 1970s and 1980s (Perrow, 2011b; Roe & Schulman, 2008, 2016; Schulman & Roe, 2018).

Schulman and Roe make this point very clearly, "in our research on modern infrastructures such as electric power, water, marine transport, telecoms, and dams and levees it became apparent that the outputs of any one of these infrastructures can depend on the inputs from others (…) Already this has led to increased social vulnerability to infrastructure failure, and not just terrorist attack. As is now well recognized, the failures of modern infrastructures can create their own social catastrophes" (Schulman & Roe, 2018, 205, 209).

Their list is obviously not exhaustive, and many infrastructures ranging from Internet, finance or aviation could be added. They are not all of them connected to global flows to the same extent, but the concept of systemic risks in relation to critical infrastructures is now well integrated in the discourse of globalisation (Goldin & Mariathasan, 2015; Centeno et al., 2015; Arnoldi, 2016; Rehgezza-Fit, 2016; Renn, 2016). Thus, for Goldin and Mariathasan, "globalisation is a double-edged sword that can be a force for progress as well as a source of great harm" (Goldin & Mariathasan, 2015, 30).

They retain and discuss the following systemic risks associated to globalisation: financial risks, supply chain risks, infrastructure risks (transportation, energy and Internet[1]), pandemics and health risks, inequality and social risk. Thus, from the early work of Beck and Giddens in the 1990s, the notion of systemic risk has finally reached the status of a standard way of addressing safety problems in an increasingly intertwined and globalised world. It is for this very reason, another category of risk along with sociotechnological risks. This new category is, however, not the only one to come out as a result of transformations and concerns of the past two decades. Another one, existential risk, connected to notions such as anthropocene and transhumanism, is discussed next.

Anthropocene, Transhumanism and Existential Risks

The following new category of risk introduced now has not yet been discussed in the preceding chapters and is linked to another scale of problems, for which the concept of anthropocene and transhumanism plays a central role. This notion of anthropocene was first proposed during an international conference on the geosphere–biosphere in 2000 by the atmospheric chemist

[1] And the expansion of the Internet infrastructure has created new "digitalised risks" including, but not limited to, cyberattacks (Lupton, 2016).

Paul Crutzen (Steffen et al., 2011; Hamilton et al., 2015; Bonneuil & Fressoz, 2016). The anthropocene is an expression that characterises a moment of our history during which humanity as a whole, as a result of its industrial activity, acts as something equivalent to a geological force. The intensity and extent of human activity on earth is sufficient to modify considerably the dynamic of our planet.

This is a scientific terminology framing the underlying causes for what is known for several years, even decades now, as global warming triggering climate change. But it does more than this by both materialistically and metaphorically assimilating humanity as a geological force. It incorporates as a result inextricably humans within their ecological habitat and adds support to an already-existing thread of critics of the nature/culture dualism (Dewey, 1935; Morin, 1977; Latour, 1991).

Two years before this proposition of anthropocene but without any link with it, in 1998, the transhumanist declaration was signed by a number of authors (Bostrom, 1998, 2005). This time, it is not humans as part of nature that propels the writers into their transhumanist declaration but rather the prospect of using new technoscientific possibilities for the advent of a posthuman. Transhumanism is the project of alleviating mankind to many of its (supposedly) biological, physiological, cognitive and social limitations (Besnier, 2010).

It is the realisation of the cyborg, an augmented human advocated through the convergence of the latest technoscientific advances such as complexity and information sciences, bioengineering, nanotechnology and neurocognitive sciences. As for the anthropocene, such a transhumanist materialisation concretely moves humanity beyond the traditional divide or dualism of a human nature existing independently of its technological milieu. It is in this context of both anthropocene and transhumanism that the ideas of collapse (Diamond, 2005; Ehrlich & Ehrlich, 2013) and existential risks (Bostrom, 2002; Bostrom & Cirkovic, 2008) have, in the past decade, translated into new risk categories that would provoke threats of very large, one could say apocalyptic (but in a secular way), scale.

The definition of existential risks is related to events with the potential to threaten the survival of humanity. In this respect, Diamond's book has done much, because of its depth, breadth, commercial success (and despite its critics), to stimulate an array of popular concern about the possibility of extinction or collapse of complex societies. By grouping into one integrated study many examples of ancient cultures such as the Maya civilisation, the Easter Island society or the Anazis in America, which disappeared at different moment of history, Diamond created a compelling argument.

> The parallels between Easter Island and the whole modern world are chillingly obvious. Thanks to globalization, international trade, jet planes, and the internet, all countries on Earth today share resources and affect each other, just as did Easter's dozen clans. Polynesian Easter

> Island was isolated in the Pacific Ocean as the Earth is today in space (...) Those are the reasons why people see the collapse of Easter Island society as a metaphor, a worst-case scenario, for what may lie ahead of us in our own future.
>
> *Diamond (2005, 119)*

Environmental damage (1), climate change (2), hostile neighbours (3), decreased support by friendly trade partners (4) and the degree and nature of society's response to its problems (5) constitutes the framework through which he analyses past ecological catastrophes that triggered societal collapses. Diamond does not promote an environmental determinism but show instead the complex relationship between these different dimensions.

But the global situation is worrisome to say the least and represents issues such as accelerating extinction of animals and plants populations and species (biodiversity), land degradation, ocean acidification and eutrophisation (dead zones), worsening of some aspects of the epidemiological environment and depletion of increasingly scarce resources (Ehrlich & Ehrlich, 2013).

This type of analysis is now central to many concrete and intellectual initiatives, which have incorporated the possibility of a collapse of our societies and suggest, develop and sometimes already implement strategies of resilience to cope with this potentiality (Cochet, 2009; Servigne & Stevens, 2015). Although Diamond's book and the notion of collapse of our societies is now fully admitted and their implications concretely explored, it is the category of existential risk rather than collapse, in my view, which embraces best these entire new scales of problems.

Bostrom defines existential risk as "one where an adverse outcome would either annihilate Earth-originating intelligent life or permanently and drastically curtail its potential" (Bostrom, 2002, 2). His idea is to address a specific category of events, which goes beyond what we traditionally associate with catastrophes such as earthquake, volcano eruption, draughts, epidemics, Chernobyl or Bhopal. For Bostrom "with the exception of species destroying comet or asteroid impact (an extremely rare occurrence), there were probably no significant existential risks in human history until the mid-twentieth century, and certainly none that it was within our power to do something about" (Bostrom, 2002).

In a collective book (Bostrom & Ćirković, 2008), many of these are analysed and discussed. They cover events related to our cosmological condition, namely the consequences of a "nearby" collapsing star, an impact of a comet or asteroid. Other "natural" extreme possibilities considered are supervolcano eruption (e.g. Yellowstone caldera) and natural pandemic.

Followed by these exogenous possibilities are the anthropic one, man-made ones. The phenomena discussed are physic experiment, climate catastrophe, doomsday war, nanotechnology, synthetic biology or machine super intelligence.[2] The latter is not only the most intriguing but also one that has attracted much attention by media recently and associated with the development of artificial intelligence (AI). AI safety research now exists with a significant amount of funding, and with already explicit design preventive strategies (Bostrom, 2014).

The risk is, basically, to create autonomous intelligences generating their own purposes that could as a result provoke unexpected existential scenario, such as finding ways to provoke a complete loss of control by humans of the (critical) informational infrastructures with far-reaching consequences.[3] All these existential risks, also called global catastrophic risks (or extreme risks, Baum & Barrett, 2015), have been for the first time addressed and turned into policy documents and reports, published by the Global Challenges Foundation in 2015 and 2016.

Widening and Complexifying the Risk Picture

This sweeping view of these risk categories might feel overwhelming and too broad to provide analytical relevance, but what is interesting is that it indicates new phases of developments in academic interest, reflecting new concerns. Moreover, these are clearly linked and overlapping but have been framed independently, at different periods, addressing different scale, scope and time frame of phenomena. This is what this chapter is about, this expansion. Schulman and Roe also advocate an extension of scope, inter-organisations and time frame for the study of high-reliability and critical infrastructures (Shulman & Roe, 2018).

They indicate not only the need to consider several kinds of risks from accidents (e.g. explosion, fires) to long-term chronical effects of chemicals in the environment (pollution) but also the need to think organisational reliability from a broader scale as an interorganisational phenomenon (several infrastructure and organisations interacting) and regulatory one. This chapter follows the same perspective of extending scale, scope and time frame but does it to encompass systemic and existential categories introduced earlier. Let's come back on the categories introduced in this chapter.

Thus, reading studies about sociotechnological risks shows that it is predominantly something that concerns a certain scope (type of threats) and scale (as a degree of relationship between the local and global). With the works of Turner, Perrow, Vaughan or Hopkins, one finds mostly descriptions and explanations that combine engineering knowledge of artefacts

[2] We could now add the geoengineering propositions of preventing global warming (Hamilton, 2016).

[3] One problem with some of the phenomena of the category of existential risks, as for instance with the risk of superintelligence, is that the border is unclear between realistic technoscientific forecasts, on the one hand, and science fiction, on the other hand. Some strongly reject the possibility of the advent in a foreseeable future of an AI consciousness that would outsmart then threaten humans (e.g., Ganascia, 2017).

(nuclear reactor, aircraft, offshore platform, pipeline, etc.) with cognitive and social behaviour of a wide range of actors (pilots, process operators, inspector, executives, managers, etc.). It is a form category, which, despite being already quite broad, is mainly centred on the coupling of artefacts and for-profit or public organisations.

With systemic risks, the range of issues and entities introduced appears to widen the scope and scale of issues and dimensions to integrate. Pandemics, terrorism or natural catastrophes seen from the perspective of globalisation entail a wider perspective than sociotechnological risks. With Beck, Giddens or Goldin and Kurtana, the reader delves into issues dealing with climatology, geology and ecology (e.g. hurricanes, earthquake), biology (e.g. viruses, bacteries) and geography (e.g. cross continental events) while also keeping the engineering and social dimensions of artefacts, actors and institutions that connect these processes at the world stage. The scope (the range of phenomena) increases as the scale too.

The category of existential risks moves now scope and scale to another level and adds an expanded time frame. The writings of Bostrom expend the range of scenarios (scope) by situating our human conditions within the cosmos and its associated threats, whether comets or asteroids, star supernovae, solar flares or supervolcano and global warming. Second, with this new scope comes a new scale of geography and consequences because existential risks are these events with the potential to trigger societies collapse, in the worse circumstances, a human extinction.

Third, the relationship with time is also different because some of the threats are ahead, in decades to come (e.g. global warming, asteroid, technological innovation), but actions need to be taken now to build, transnationally, the capabilities to prevent or mitigate potential consequences. Fourth, it incorporates the notions of anthropocene and transhumanism with their associated extreme catastrophic potentialities, such as societies collapse or unfriendly superintelligence, which marks an ontological rupture with a dualist philosophy by assimilating humans as a geological force and by programming or prophesising the cyborg, the advent of a hybrid. Scope, scale and time frame are this time extended further in comparison with the two previous categories (Table 5.1).

Of course, this presentation contains certain simplification, two of them are now briefly commented. First, authors in the different risk categories partly overlap. This is not a sharp distinction. Perrow already introduced presystemic or pre-existential risks thinking in his writings, for instance, when discussing the international side of aviation, when characterising ecosystem accidents (more about this later) and identifying DNA engineering as a new type of high-risk systems in 1984 (Perrow, 1984). So, one option could be to see in Perrow an early writer on systemic risks, something that his later writings illustrate (Perrow, 2011b). But if Perrow was a pioneer clearly shaping the debates on sociotechnological risks, he was not really on systemic (or existential) risks that he never addressed specifically as a category.

TABLE 5.1

Expanding of Scope, Scale and Time Frame

Period	Categories of Risks	Associated Key Concept	Examples	Authors (Indicative)
2010	Existential risks	Anthropocene and transhumanism	Asteroid impact Ecological catastrophe and society collapse Supervolcano	Bostrom (2014) Bostrom and Circovik (2008) Diamond (2005)
2000	Systemic risks	Globalisation	Terrorism Pandemics Finance crisis Natural disasters	Beck (1996, 2002) Giddens (1990, 2000) Perrow (2011b) Goldin and Mariathasan (2015)
1990 1980	Sociotechnological risks	High-risk systems and large technical systems	Nuclear accident Chemical explosion Aircraft crash Train derailment	Turner (1978) Perrow (1984) Vaughan (1996) Hopkins (2000, 2012)

Similarly, Beck mentioned from time to time in his texts and very early on, for instance, in an article (Beck, 1992), the possibility of massive ecological degradation triggering potential societies collapse. He therefore could be seen as anticipating some of the aspects of the category of existential risks and therefore be considered as an early writer on this topic. But Beck was one of the first with Giddens to explicitly formulate the problem of systemic risks because of his association of risk together with globalisation, and he did not really formulate the existential risks category.

Second, sociotechnological, systemic and existential risks have longer histories than the past decades, beyond the 1980s, 1990s and 2000s. Environment and technology historians (e.g. Le Roux, 2011; Fressoz, 2012; Jarrige, 2016) have now shown how technological accidents and alerts about the ecological impacts of industrialism have long been a concern for societies. These concerns have been expressed for more than 200 years of interaction between civil society, state, industry and science on these topics.

As these authors assert, past societies did not enter modernity without any controversies about the consequences of the industrial revolution, namely about the negative side of modernity. Moreover, one could say that natural disasters and their study are far from new. Avalanches, floods, earthquakes or volcano eruptions have led to the development of an expertise in these domains for quite some time. However, natural catastrophes are more frequent in relation to global warming. A similar reasoning can be applied to terrorism. Terrorist attacks are far from new, but globalisation produces new possibilities of coordinating actions at the global stage, in particular with the help of both Internet and technologies available to build weapons for these attacks.

With pandemics, an obvious example of systemic risk in the far past is, for instance, how Europeans' germs decimated South American population in the 15th and 16th centuries (Diamond, 1997), but they now not only take an increased level of speed and reach through air transport but also potentially change of nature, as synthetic biology and bioengineering might produce new kind of viral infection. There would be, in this vein, many other examples of "preglobalisation" (as defined today) systemic risks. And, we could add that we don't discover either the notion of catastrophe, collapse or apocalypse (Hughes, 2008), which has a long history from myths and religions to secular science fiction novels and different philosophies, including ecology, rooted in the reflection of the hubris of science and technology (e.g. Ellul, 1954; Jonas, 1984).

The "nuclear winter" threat was at the heart of these early considerations of existential risks (Serres, 2015). But, although not entirely new in this sense, these categories of risks only came to be named and made explicit that way in recently. The same can be said for the macroconcept associated with these categories such as high-risk systems (Fressoz, 2012), globalisation (Berger, 2003), anthropocene (Grienvald, 2012) or transhumanism (Atlan & Droit, 2014). They, too, have longer histories than implied earlier.

The choices made in this chapter are related to the conceptual, scientific and intellectual discourses, which are built and developed into specific investigations turned in visible articles and books and then research programs, traditions and communities to provide analytical lenses in order to help understand, explain, anticipate, manage and regulate risks. It does not seek to establish the long histories that are behind each of these categories, although this is an interesting task.

What these three explicit categories of risks do is to formally translate a higher level of consciousness about the consequences of the profound successive transformations of our technosocieties, world and planet in the past 30 years. Sociotechnological risks characterise this moment during which an array of safety-critical and large technical systems expands (e.g. aviation, nuclear power plants) while requiring a better understanding of their complexities, and similarly so for systemic and existential risks. Systemic risks emerge as a concept in relation to globalisation and its consequences with a combination of threats on critical infrastructures and the ecology, and existential risks with a moment of acceleration in both technoscience hubris and ecological degradation, and consciousness of how precious and precarious life on our planet is.

Embedded Risk Categories: Complex Interactions

Although they have different conceptual histories and genesis, none of these categories of risks are disconnected but are fully embedded instead, and it is integral to the *Post NA* discourse. Sociotechnological risks (high-risk systems) are embedded in systemic risks (globalisation), which are in turn embedded in existential ones (anthropocene and transhumanism). I propose in what follows a reading of some of the puzzles associated with this nested nature of issues and the interactions between the three categories of risks in three distinct parts considering in turn.

As described in the retrospective earlier, sociotechnological, systemic and existential risks have been addressed and conceptualised at different moments in time, with their own and sometimes shared historical roots. But this of course does not mean that once systemic or existential risks are formulated, the category of sociotechnological risks subsequently disappears.

The categories remain of course valuable even when new ones appear. Sociotechnological risks such as a nuclear reactor meltdown, a chemical explosion, an aircraft crash or a train derailment exist along systemic risks such as terrorism, pandemics, finance crisis or natural disasters as well as existential risks such as asteroid impact, ecological catastrophes or unfriendly super intelligence.

However, the interactions between the three categories are highly complex involving spatial, time and networked circular causalities based on numerous (and partly unpredictable) feedback and feedforward loops over the past decades and in the next years to come. The scope, scale and time frame of our grasp of high-risk systems is now modified. In order to simplify the picture, the following visualisation (Figure 5.1) is commented next at a macrolevel of description on the basis of the two loops A and B of the representation, divided into four interrelated sequences.

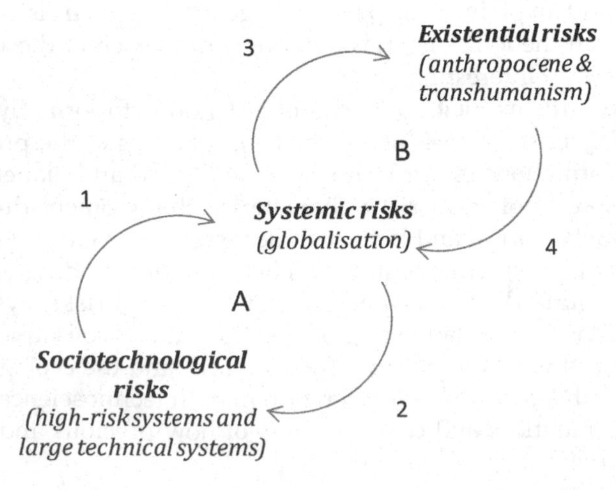

FIGURE 5.1
Complex relationships between risk categories.

Loop A: Causal Circularity between High-Risk Systems and Globalisation

1. High-risk and large technical systems have become the basic infrastructures of our global world. Trains, aircrafts, electricity, boats, oil and gas, food supply distribution and the cables, and satellites of the information (or cyber) infrastructure are the networks on which globalised processes currently expand and thrive. Their day-to-day operations are not only needed to the lifestyle of a growing amount of people but also expose employees, users and third parties to certain kind of risks. Globalisation is intrinsically connected to these high-risk and large technical systems that have become the critical infrastructures of our epoch. Some high-risk and large technical systems are early examples of internationally operated activities requiring a supranational level of regulation and coordination in order to ensure safety and reliability at an acceptable level. Two prime cases are civil aviation and nuclear industry which developed international organisations (ICAO and IAEA) at different moment of their histories. So, to study globalisation and its associated systemic risks requires investigating the functioning of sociotechnological systems at different scales, from national to supranational levels.
2. But globalisation retroactively impacts these systems in at least three different ways. First, systemic risks such as terrorism, natural events or financial crisis represent threats to their integrity and must now be integral part of their safety management and governance. These infrastructures are sometimes both targets and weapons of terrorism (e.g. aircrafts as in 9/11) but are also part of the problem of global warming (e.g. transportation systems CO_2 emission), which retroactively impact other infrastructures with the increase of natural events (e.g. earthquake, hurricanes; Fukushima Daïchi is one example, to which I turn to later). Second, their degree of interconnectedness creates problems when one of these critical infrastructures is impaired for internal or external reasons, with ripple effects throughout the networks (e.g. the ashes cloud from the Iceland volcano in 2014 is an example[4]). Third, globalisation pushed by privatisation, liberalisation of finance and trade, ICT and transport revolutions has completely modified the operating conditions and landscapes of high-risk systems, large technical systems and critical infrastructures. Examples of trends deriving from globalised processes are standardisation, externalisation, financialisation or self-regulation which have shaped a configuration of polycentric

[4] Events are also global because they are broadcasted live everywhere in the world, and are transnational in the sense that they can affect people from everywhere in the world because of world flows, including tourism. Sweden suffered great losses from the tsunami in Southeast Asia as a result of 100 citizens on holidays in Thailand at the time (Reghezza-Zitt, 2016).

dynamic of a multitude of artefacts, entities, actors, organisations and institutions. This has considerably modified the condition for reliability, safety and resilience of these systems (this point was discussed in Chapter 4).

Loop B: Causal Circularity between Globalisation and Anthropocene/Transhumanism

3. In relation to the anthropocene, globalisation, in its industrialised and neoliberal side, currently pushes the development of many nations across the world and has become a key dynamic that contributes to global warming. The need of many nations to rely on similar infrastructures than the developed western world creates many challenges for the ambitions of reducing the phenomenon of global warming. Second, by promoting technological innovations in a highly competitive world with limited democratic or regulatory controls at national and transnational levels (technological innovation is led by powerful globalised multinationals, and a philosophy of progress through technology), globalisation is likely to fuel experiments that promise to deliver competitive advantage to nations (e.g. nanotechnology, synthetic biology, AI), exposing employees, users and third parties to potential risks, including those described in the category of existential risks. And the materialisation of these new sociotechnological systems based on superintelligence, nanotechnology or bioengineering could be of unprecedented scale, scope and dynamics, questioning the conditions under which they remain reliable, safe or resilient.
4. Finally, the more than ever plausible anticipation of a collapse has considerable potential to curtail the existence of current high-risk, large technical systems and critical infrastructures. Scenarios of doom have propelled advocates to promote a radical redesign of our modes of living. The development of local sustainability that relies on supply from renewable energy, proximate food circuits and less consumerist lifestyles would mean the end of the society as we know it and also reduce the existence of some of our current infrastructures including existing high-risk systems (nuclear power plants, refineries, offshore platforms). Conversely, new ecological modes of producing also have their own associated sociotechnological risks to be anticipated and managed (e.g. hydrogen, CO_2 storage, etc.). Existential risks derived from sociotechnological developments involving AI or biotechnology also trigger concerns that should push world leaders and civil societies to design transnational responses, which would involve redesign of current institutions and critical infrastructures.

Rethinking Perrow's Matrix

Scale of Governance and Magnitude of Impact

Altogether I believe that these broad reflections about the contemporary situation convey a different perspective than the original Perrow's 2 × 2 matrix, a different perspective that might be interesting to visualise. As a start and following Chapters 1 and 2, one can say today that there is more than interactivity (linear or complex) or coupling (tight or loose) to be considered when one studies disasters, and a more managerially, socially, strategically and regulatory oriented analysis is needed. We now know that major accidents happen whether or not systems are tightly coupled or interactively complex, and for reasons that are not explained by coupling and interactions but other ones discussed in previous chapters.

As a consequence, these two axes (coupling, interaction) are not so central anymore because they cannot help predict disasters in relation to a position of system in the 2 × 2 matrix. And different dimensions are required to reflect what has been said in this chapter and the others. These two axes can therefore be abandoned and replaced to update Perrow's original visualisation. Then what could we substitute to replace these continuums of coupling and interaction? Here is a proposition that reflects the content of this chapter while rethinking Perrow's original matrix.

First, we can notice that the categories of risks established in the last 10–20 years in parallel with the more familiar notion of sociotechnological disasters create a continuum that can be defined as "magnitude of impact" which ranges from major to extreme risks. Second, because of the globalised nature of many of these risks, the notion of "scale of governance" becomes an important category not only when considering the issue of preventing them but also when it comes to describing their modes of operating. Considering these two options of magnitude of impact and scale of governance, an alternative proposition to Perrow's matrix of risk is illustrated in Figure 5.2.

This graphics translates a very different set of questions and also landscape which were not at all integral of *NA*, simply because more than three decades have passed and Perrow's questions about high-risk systems cannot be formulated the same way. The principles of Perrow's 2 × 2 matrix are surely replaced completely, and the new graphics is a very different one. But it operates in a similar fashion than Perrow's one in *NA*, using the heuristic value of visualisation to complement sentences of the *Post NA* discourse.

The attempt to visualise in a spatial rather than in a sequential manner, as the text earlier provides a valuable new angle to contemplate together these categories, and their relations, along with Figure 5.1. about the complex relationships between these categories. Clearly the prevention of a train

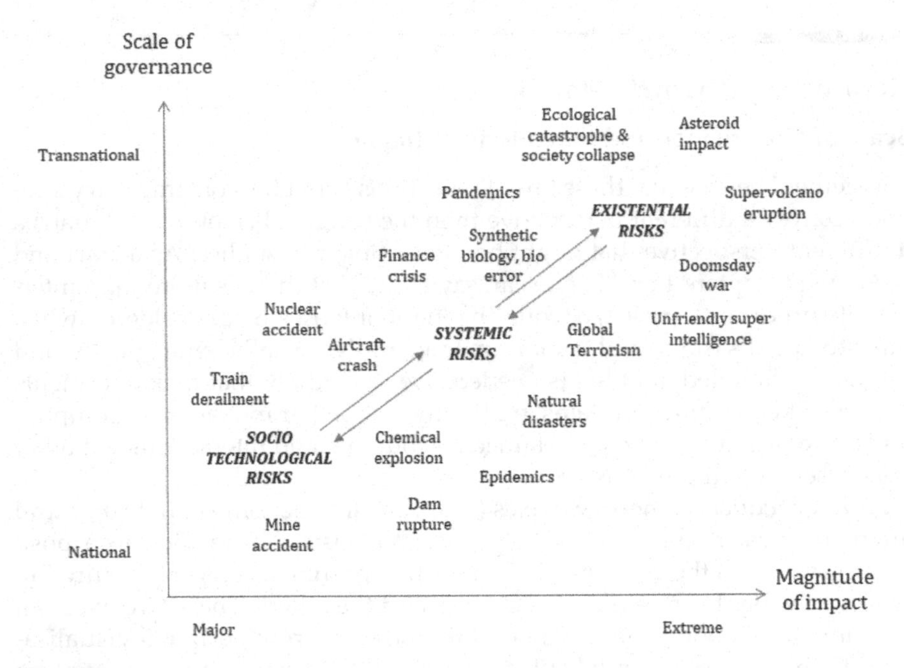

FIGURE 5.2
Risk categories in relation to scale of governance and magnitude of impact.

derailment or mine explosion does not require the same level of transnational governance than the prevention of an ecological catastrophe, and its potential magnitude of impact is obviously not the same. The same applies to a dam rupture and an asteroid impact.

But these different categories of risk are not independent, and they interact in the ways briefly elaborated and commented previously through different loops in Figure 5.1 and visually indicated with the help of arrows in Figure 5.2. For instance, a natural event (storm) increased by global warming can impact locally a train or a dam (e.g. Katrina, 2005), and a tsunami, a nuclear plant (e.g. Fukushima, 2011). Another example is the Internet infrastructure that creates cybersecurity threats.

It is worth emphasising again that the interactions between the three categories of risks are highly complex involving spatial, time and networked circular causalities based on numerous (and partly unpredictable) feedback and feedforward loops over the past decades and in the next years and decades to come. Natech (for natural hazards causing technological disasters) and cybersecurity are two examples of how systemic and existential risk categories currently translate in safety management concerns. This is a major difference between *NA* and *Post NA* for which scope, scale and time have expanded and complexified.

Eco-Socio-Technological Disasters

The Case of Fukushima Daïchi

If the BP story was the paradigmatic case exemplifying how globalisation created a new landscape for high-risk systems (Chapter 4), Fukushima Daïchi nuclear reactor meltdown could well be the paradigmatic example of this new picture, captured in Figure 5.2. The events in Japan in 2011 are well known. The meltdown happened because of an earthquake which became a tsunami which then turned into a devastating force defeating the engineered systems of the Fukushima Daïchi nuclear power plant.

If one sticks to the analytical tools of na in 1984, such as "tight coupling" and "interactive complexity", Fukushima Daïchi can be described well. Perrow developed the notions of "tight coupling" and "interactive complexity" for the internal features of high-risk systems and from a socioengineering perspective. It was indeed triggered by his interpretation of Three Mile Island (Perrow, 1982).

It defined a type of interactions that were not unexpected in the context of the na thesis. Perrow targeted these interactions that were "not designed into the system by anybody; no one intended them to be linked" (Perrow, 1984, 75). These interactions can become complex when "there are branching path, feedback loops, jumps from one linear sequence to another because of proximity and certain other features" (Perrow, 1984, 75).

But, although predominantly elaborated with man-made system in mind, Perrow had also anticipated the concept of "ecosystem accident". He described it this way "eco-system accidents illustrate the tight coupling between human made systems and natural systems. There are few or no deliberate buffers inserted between the two systems because the designers never expected them to be connected" (Perrow, 1999, 296). So, one could conclude that this works well with Fukushima Daïchi. Designers did not anticipate the link between the plant and the tsunami.

The story is of course different and requires, again, the second thesis of *NA*. Indeed, earthquake and tsunami were both anticipated, and design choices made (Susuki & Nakai, 2012). However, the dam protecting the plant was too low to contain the size of the wave, despite existing assessments of its potential height limitation against which the company, Tepco, argued, such knowledge being always ambiguous and uncertain (Downer, 2014). But, most likely, the cost of building a higher one slowed down its construction.

Moreover, the backup batteries were useless because of their common failure mode. They were all located below the water level, a problematic design choice in case of flooding. Finally, a failsafe system designed to shut the condenser valve in case of loss of energy impeded the self-cooling of the reactor, although it is precisely what was needed. Because the engineers and operators in the control room did not know about this failsafe function, they couldn't do everything they could, and the reactors heated up then melted down. These were design and operating flaws that transformed a tsunami into a nuclear disaster.

Again, the second thesis of *NA* proves far more effective to understand this event. This is precisely why Perrow commented Fukushima Daïchi this way, a quote already used in Chapter 1. "Nothing is perfect, no matter how hard people try to make things work, and in the industrial arena there will always be failures of design, components, or procedures. There will always be operator errors and unexpected environmental conditions. Because of the inevitability of these failures and because there are often economic incentives for business not to try very hard to play it safe, government regulates risky systems in an attempt to make them less so" (Perrow, 2011a, 44).

So Fukushima Daïchi could be a mix of the second thesis and of an "ecosystem accident", but the distinctions between natural, biological, ecological and social realities remains engrained in Perrow's sociology. In his version of an ecosystem accident, one understands that there are "human-made systems" on one side and "natural systems" on the other. Pritchard, an historian of environment and technology commenting on Fukushima insists more than Perrow does on the hybrids of nature–culture precisely for this reason. "Humans are part of the natural world, even, we might add, in highly managed, engineered technological spaces like those of nuclear reactors" (Pritchard, 2012, 229). Her proposition is to complement the na framework through the notion of "envirotechnical disaster" (Pritchard, 2012).

Pritchard relies on the concept of hybridity to anchor her argument of a new version of na beyond the ecosystem accident. "Hybrids of nature-culture and nature-technology surround us. And arguably, we humans are hybrid, too. Fukustima may not only confirm this insight but even come to illustrate them in ways that are powerful, disturbing, and also humbling" (Pritchard, 2012, 222). What she has in mind is the more profound modification of our ways of thinking humanity following writers such as Latour and his concepts of hybridity, of network (Latour, 1991). Another possibility suggested in this chapter is to keep Perrow's notion of complexity while specifying a new meaning for this concept (Morin, 1977, 2007).

General Complexity

Complexity has indeed several meanings and several decades of history (including cybernetics in the 1950s) but attracted more attention in the 1980s with the publicised activity of the Santa Fe Institute (e.g. Waldrop, 1992; Lewin, 1992; Gell-Man, 1995; Mitchell, 2009; Johnson, 2009). The basic ingredients of complexity thinking are now widely shared and translated in different disciplinary contexts. These ingredients are positive and negative feedback loops, circular and nonlinear causality, the interaction of order and disorder, the sensitivity to initial conditions and the problem of determinism, the properties of emergence and self-organisation of complex adaptive systems and also the issue of decomposition of systems and reductionism.

But it is important to distinguish the complexity of computer simulations, of the idea that it is still possible to produce objective laws which will help determine the course of events in complex systems. "Most complexity researchers have not yet reflected about the philosophical foundations of their approach (…) As such, many still implicitly cling to the Newtonian paradigm, hoping to discover mathematically formulated 'laws of complexity' that would restore some form of absolute order or determinism to the very uncertain world they are trying to understand" (Heylinghen, Cillers, & Gershenson, 2007).

In contrast, other complexity perspectives break this mould. They oppose this restricted complexity which remains anchored in the popular account of Cartesian and Newtonian philosophies, to a general material, ecological, epistemological and philosophical complexity that offers a mode of thinking which is more apt to guide our attempts to develop alternative mindsets in tune with the current challenges of our epoch, as for instance, described by the complex interactions described earlier between sociotechnical, systemic and existential risks (Figure 13) and illustrated by Fukushima Daïchi disaster. It is the philosopher Morin who suggests differentiating restricted from general complexity, borrowing the expressions restricted and general from Einstein's restricted versus general relativity (Morin, 2007).

Relying on the key notion of causal circularity or feedback causation, general complexity promotes the very long historical continuity as established by the diversity of sciences (but without reductionist ambition) between humans, societies and their geological, biological and ecological processes, through a series of self-organised emergences in the course of earth, biological and then human history.

In general complexity, the notion of emergence is connected to both the limits of reductionism and to the notion of event. Reductionism is the idea that one can break down matter to obtain the ultimate brick of our world from which one can then ideally infer higher levels of phenomena, such as deducing society from minds, minds from neurons, neurons from chemical processes and chemical processes from physical ones. Despite the opposite rhetoric, this ideal is found in the restricted complexity view which often expects computers to provide the simulation to move from the micro to the macro with some serious limitations not always reflected upon (Jensen, 2018).

General complexity through concepts of self-organisation, causal circularity and emergence advocates irreducible levels of phenomena that cannot be inferred from their constitutive parts. One can talk of degrees of emergence as advocated by Deacon, which characterises the combination of a temporal dimension (i.e. diachronic) with a hierarchical or network one (i.e. synchronic) to distinguish different nature of emergences (Deacon, 2003). One understands that the mind emerges from the brain in a very different way than water emerges from oxygen and hydrogen. Complexity in these two cases is not of the same degree.

Emergence also means potential for radical novelty which cannot be predicted, a creativity expressed by processes making up our world and defying predictions. So, by substituting nonlinear causality to linear causality while assembling natural, biological, ecological and social entities together without reducing one level to another but by conceptualising their continuous interactions and intertwined dynamic, general complexity opens our mode of investigation to a kind of ecological sensitivity and greater appreciation of how intricate the texture of the world is. It is this time the Cartesian man "as master and possessor of nature" which is challenged.

General complexity can therefore accommodate with a range of disciplines that have flourished in the past decades against a posture of anthropocentrism, of a human (and society) out of nature. Disciplines such as political ecology, environmental law and ethics, environmental history, anthropology of nature, ecological economics, environmental philosophy, sociology of reflexive modernity and actor–network theory have promoted an ecological worldview in social sciences (Debourdeau, 2013).

So general complexity goes beyond the "complexity turn" of the social sciences advocated by a sociologist such as Urry (2005) by reconsidering some of the core assumptions of sociology still present, for instance, in Perrow's writings (or also in Hopkins' and Vaughan's) which remain within the anthropocentric view. The most widespread terminologies to conceptualise this reformatting of the social are *nonhumans, hybrids* and *networks*, advocated by Latour during the 1980s and 1990s (e.g. Latour, 1991) and incorporated in social sciences research (Houdart & Thiery, 2011).

This is what Pritchard advocates (Pritchard, 2012, see earlier). But turning to complexity with this new meaning is another option while keeping a link to this important concept used by Perrow. Seen from general complexity, Fukushima Daïchi is the failure of engineers, managers and regulators to embrace the full implications of designing artefacts which are inextricably embedded in nature, and not outside nature. To follow Pritchard but slightly modifying her proposition of terminology of envirotechnical disaster, Fukushima could be described as an eco-socio-technological disaster, which leads to a modification of the figure depicting the move from *NA* to *Post NA* (Figure 5.3). In this new figure based on Figure 4.2, one more layer is now added to signify the importance of ecosystems that are no longer considered to be peripheral and local, as they were some decades ago but fundamental and global in our current context.

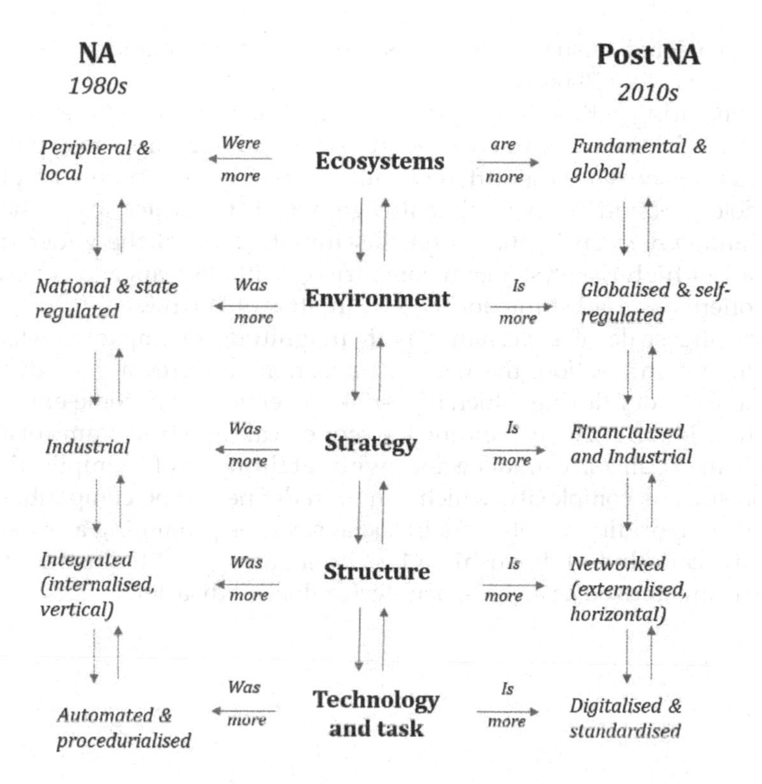

FIGURE 5.3
From *NA* to *Post NA*, eco-socio-technological.

Summary of Chapter 5

Since the publication of *NA*, the world has witnessed a great number of changes that have affected our perception, understanding and conceptualisation of high-risk systems. One way to characterise these mutations is to keep track of new categories of risks which have been produced by scholars in the past two to three decades. When one does so, the notion of systemic risks appears to be concomitant with the description by authors of the phenomenon of globalisation in the 1990s.

Increase of interdependencies across the world through a diversity of flows creates possibilities of risks of new scope and scale in comparison with sociotechnological risks created to frame the advent of high-risk systems in the 1980s. The recent category of existential risks can be linked to concerns for potential collapse of society due to a variety of man-made and anthropic events, connected to such concepts as the anthropocene and transhumanism, products of the 2000s.

Overwhelming as this picture can be, it constitutes the new context of high-risk systems for the years and decades to come, and our ability to explore the interactions between these different categories of risks is a new empirical, methodological and conceptual challenge, which translates topics such as natech and cybersecurity. Based on this investigation of the wider trends which affect high-risk systems in comparison with 36 years ago, a new picture is offered as a substitute for the 2 × 2 matrix of Perrow.

Combining scale of governance with magnitude of impacts instead of coupling and interaction, the new visualisation captures a very different moment of history during which high-risk systems have become embedded in much wider concerns, expanding the scope, scale and time frame of analysis. Such an expansion cannot be done without thinking of its implication for a notion such as complexity, which can be redefined to be compatible with concepts of hybridity or networks in social sciences promoting a revision of our analytical mindset. Fukushima Daïchi, a paradigmatic illustration, can then be characterised as an eco-socio-technological disaster.

6

Conclusion

Post Normal Accident (*Post NA*) provides in the early 2020s a new mindset adapted to the conceptual and empirical changes of the past decades, which followed the publication of the book *Normal Accident* (*NA*), by Perrow, in the 1980s. It provides a set of analytical tools to help study and sensitise empirically high-risk systems whether retrospectively, after major event, or in daily operations. The intention of this new discourse or narrative based on these several articulated analytical lenses, exposed in the chapters of this book and now briefly summarised in this conclusion, is indeed to offer a broad perspective, within a single and coherent picture (Table 6.1).

TABLE 6.1

Normal Accident and Post Normal Accident

Normal Accident	***Post Normal Accident***
The book *Normal Accident* (*NA*) is about the normal accident (na) thesis	The book *Normal Accident* (*NA*) also contains the opposite thesis to normal accident (na)
Some Accidents are normal because of interactive complexity and tight coupling	Accidents are normal because they repeat despite knowing that they could be prevented in principle
Human error of front-line actors is not a good explanation for disasters	Strategic error of top management is a better explanation for disasters
Component failure accidents	Network failure accidents
Technological accidents *e.g. Piper Alpha, Chernobyl*	(Global) Eco-socio-technological accidents *e.g. Deepwater Horizon, Fukushima Daïchi*
Matrix combining "coupling" and "interaction"	Graph combining "scale of governance" and "magnitude of impact"
Interactive complexity *Technological risks*	General complexity *Sociotechnological, systemic and existential risks*

Two Opposite Theses in One Book

First, *Post NA* argues that *NA*, the book, must be read differently than the traditional way. *NA* contains the thesis that some accidents are inevitable (normal accident, na), but that is not all. *NA* contains also another second, completely opposite, thesis, based on a power view of organisations. *Post NA* asserts that this opposite thesis considers *accidents to be normal because they repeat despite knowing that they could be prevented in principle.* This view is supported by a historical perspective of the work of Perrow over the years and by situating *NA* as a book illustrating negative externalities, technological disasters and, in certain kind of businesses, high-risk systems.

Triggered by Three Miles Island in 1979, his analysis goes far beyond the technological determinism associated with the book to embrace a more organisational, social, political but also critical analysis of such high-risk systems. This assertion becomes less surprising when one knows Perrow's sociology which combines technology, structure, goal and environment of organisations into a broad and critical approach of the society of organisations. By revealing this background, the second thesis appears to be as interesting, perhaps even more, than the first thesis.

Hopkins, an Unofficial Theorist for the Second Thesis of *Normal Accident*

Second, this posture is also well developed by Hopkins' sociological work, another sociologist who specialised in safety and the study of disasters. Although one could see in Hopkins an opponent of Perrow because of his explicit rejection of the main thesis in *NA* in the late 1990s and early 2000s, his work turns out, retrospectively, to be highly compatible with Perrow's second thesis. Over the years, Hopkins has produced persuasive and practical accounts of disasters which conceptualise safety through a combination of appropriate risk awareness, rule management, indicators, incentives, structure, culture, learning, auditing, (mindful) leadership and environment (industrial, legal, financial and judicial).

It perfectly supports Perrow's assertion that most accidents can be prevented if enough attention, resources and expertise are granted by companies (and by regulations) to safety. Both sociologists share a power view of safety in high-risk systems and have, as a result, also a critical view of the notion of (safety) culture. In *Post NA*, the position arguing that most accidents are preventable if enough attention is granted by top management to the issue of safety pursues the "high-reliability organisation versus na" debate of the 1990s. It however slightly differs because of its critical tone (emphasising power over culture), its distance with the technological argument (it is not tight coupling and complex interactions which cause major problems), its broader empirical and analytical scope (from shop floor to political economy) and its explicit insistence on the importance of top management in the safety of high-risk systems.

Errors from the Top (versus Sharp-End Human Errors)

Third, *Post NA* asserts that it is important to follow this new version of the debate and that it is fruitful to migrate from shop floor to top decision-makers when it comes to understanding both safety and disasters. Perrow excluded human errors of front-line actors as an explanation for disaster but implied that errors from the top were far more relevant. Whereas it clearly is a simplification to stop at the level of front-line human error, the inclusion of errors from the top of organisations offers indeed a far more heuristic and deeper angle of analysis to understand the genesis of disasters (as much as the sources of high safety levels).

In this respect, strategic decision-making studies of top management offer invaluable insights to our grasp of how it is that some high-risk systems sometimes fail and some don't. Strategy in relation to safety can be defined as "choices made by top managers over the main orientation of a business in its market and the ability of these orientations to be successful without compromising safety considering the human and technical capabilities involved". However, a focus on strategy must not become a new hindsight bias or a reductionist account of disasters (and safety) but a most needed analytical principle instead, to make sense of situations.

Network Failure Accidents (versus Component Failure Accidents)

Fourth, *Post NA* reconsiders the value of the notion of component failure accident developed by Perrow following the important transformations of the landscape of high-risk systems in the past 30 years. Globalisation has been propelled two to three decades ago by a combination of liberalisation of trade and finance, privatisation and deregulation but also two technological revolutions: one of information and communication and the other of transport (maritime, aviation).

Globalisation can be decomposed in trends such as externalisation, financialisation, standardisation, digitalisation and self-regulation, which have reconfigured safety critical systems into networked ones. The suggestion is, considering these powerful and pervasive trends, to replace component by networks into the concept of network failure accidents. BP provides a vivid illustration of what is implied in this new expression.

From Technical to Eco-Socio-Technological Disasters

Fifth, *Post NA* sees these globalised processes which have modified the operating landscape as one aspect of an expansion of the scope, scale and time frame to be considered when studying high-risk systems since the publication of *NA*, 36 years ago. Captured by notion such as systemic and existential risks, this scope, scale and time frame expansion modifies our perception of the current challenges of maintaining and ensuring safe performance of high-risk systems, as for instance, indicated by mounting concerns over natech (natural hazards triggering technological accidents) and cybersecurity. This shift leads to revise Perrow's 2×2 matrix of "coupling" and "interaction" by a graph combining "scale of governance" with "magnitude of impact."

As a consequence of this expansion, interactive complexity as conceptualised by Perrow in *NA* becomes a general complexity in *Post NA*. General complexity contests the distinctions between nature and culture and embraces the circular causalities connecting humans within ecosystems, with profound implications in the way high-risk systems are conceptualised, understood and designed. This shift is captured by the notion of eco-socio-technological disasters (or safety), which the Fukushima Daïchi event provides a paradigmatic example.

References

Ackroyd, J., Murphy, S. (2013). Transnational corporations, socio-economic change and recurrent crisis. *Critical Perspective on International Business*. 9 (4). 336–357.

Almklov, P. G., Lamvik, G. M. (2018). Taming a globalized industry–Forces and counter forces influencing maritime safety. *Marine Policy*. 96. 175–183.

Almklov, P., Antonsen, S. (2019). Digitalisation and standardisation: Changes in work as imagined and what this means for safety science. In Le Coze, J. C. (ed). *Safety Science Research: Evolution, Challenges and New Directions*. Boca Raton, FL: CRC Press, Taylor & Francis Group. 3–19.

Almklov, P., Rosness, R., Storkersen, K. (2014). When safety science meets the practitioners: Does safety science contribute to marginalization of practical knowledge? *Safety Science*. 57. 25–36.

Alvesson, M., Thompson, P. (2006). Post bureaucracy? In Ackroyd, S., Batt, R., Thompson, P., Tolbert, P. S. (eds). *The Oxford Handbook of Work and Organization*. New York: Oxford University Press.

Antonsen, S. (2020). Revisiting the issue of power in safety research. In Le Coze, J. C. (ed). *Safety Science Research: Evolution, Challenges and New Directions*. Boca Raton, FL: CRC Press, Taylor & Francis Group.

Appadurai, A. (1996). *Modernity at Large: The Culture Consequences of Globalisation*. Minneapolis, MN: University of Minnesota Press.

Appadurai, A. (2006). *Fear of Small Number: Essay on the Geography of Anger*. Durham, NC: Duke University Press.

Arbesman, S. (2016). *Overcomplicated: Technology at the Limits of Comprehension*. New York: Penguin Random House.

Arnoldi, J. (2016). Global risk. In Burgess, A., Alemmano, A., Zin, O. (eds). *Routledge Handbook of Risk Studies*. London: Routledge.

Atlan, M., Droit, R. P. (2013). *Humains: Une enquête philosophiques sur ces révolutions qui changent nos vies*. Paris: Flammarion.

Auerswald, P. E., Branscomb, L. W., La Porte, T. M., Michel-Kerjan, E. O. (2006). *Seeds of Disaster, Roots of Response: How Private Action Can Reduce Public Vulnerability*. New York: Cambridge University Press.

Banks, V. A., Plant, K. L., Stanton, N. A. (2018). Driver error or designer error: Using the cycle model to explore the circumstances surrounding the fatal Tesla crash on 7th May 2016. *Safety Science*. 108. 278–285.

Baum, S. D., Barrett, A. M. (2015). The most extreme risks: Global catastrophes. In Bier, V. (ed). *The Gower Handbook of Extreme Risks*. Farnham, UK: Gower.

Baumard, P., Starbuck, W. H. (2005). Learning from failure: Why it may not happen? *Long Range Planning*. 38. 281–298.

Beck, U. (1992). *Risk society: Towards a new modernity*. London: Sage Publication.

Beck, U. (1996). World risk society as cosmopolitan society? Ecological questions in a framework of manufactured uncertainties. *Theory, Culture and Society*. 13 (4). 1–32.

Beck, U. (2002). The terrorist threat. World risk society revisited. *Theory, Culture and Society*. 13 (4), 19 (4). 39–55.

Berger, S. (2003). *Notre première mondialisation: Leçon d'un échec oublié*. Paris: Seuil/La République des idées.
Berger, S. (2005). *How We Compete: What Companies Around the World Are Doing to Make It in the Global Economy*. New York: Doubleday.
Bergin, T. (2012). *Spin and Spill: The Inside Story of BP*. London: Random House Business.
Besnier, J. M. (2010). *Demain, les posthumains*. Paris: Le Seuil.
Bierly, P. E., Spender, J. C. (1995) Culture and high reliability organizations: The case of the nuclear submarine. *Journal of Management*. 21 (4). 639–656.
Birch, K., Peacock, M., Wellen, R., Hossein, C., Scott, S., Salazar, A. (2016). *Business and Society: A Critical Introduction*. London: Zed Books.
Bonneuil, C., Fressoz, J.-B. (2016). *L'événement anthropocène: L'histoire, la terre et nous*. Paris: Le Seuil.
Bostrom, N. (1998). The Transhumanist Declaration. http://www.transhumanism.com/declaration.htm.
Bostrom, N. (2002). Existential risks. *Journal of Evolution and Technology*. 9 (1). 1–31.
Bostrom, N. (2002). Existential risks. Analyzing human extinction scenarios and related hazards. *Journal of Evolution and Technology*. 9 (1).
Bostrom, N. (2005). A history of transhumanist thoughts. *Journal of Evolution and Technology*. 14 (1).
Bostrom, N. (2014). *Superintelligence: Path, Danger, Strategy*. Oxford, UK: Oxford University Press.
Bostrom, N., Ćirković, M. M. (eds) (2008). *Global Catastrophic Risks*. Oxford, UK: Oxford University Press.
Bourrier, M. (1999). *Le nucléaire à l'épreuve de l'organisation*. Paris: Presses Universitaires de France.
Bridge, G. (2008). Global production networks and the extractive sector: Governing resource-based development. *Journal of Economic Geography*. 8. 389–419.
Browne, J. (2010). *Beyond Business: An Inspirational Memoir from a Visionary Leader*. London: Weidenfeld & Nicolson.
Brunsson, N., Jacobson, B., Associates. (2002). *World of Standards*. Oxford, UK: Oxford University Press.
Buck, T., Buchan, D. (2002). The Sun King of the Oil Industry. *Financial Times*. July.
Burgelman, R. A., Grove, A. S. (1996). Strategic dissonance. *California Review Management*. 38 (2). 8–28.
Busch, L. (2011). *Standards: Recipes for Reality*. Cambridge, MA and London: The MIT Press.
Cardon, D. (2019). *La culture numérique*. Paris: Presses de Science Po.
Carpenter, D., Moss, D. A. (eds) (2014). *Preventing Regulatory Capture: Special Interest Influence and How to Influence It*. New York: Cambridge University Press.
Carroll, P. B., Chunka, M. (2008). *Billion Dollar Lessons: Learn from the Most Inexcusable Business Failures*. New York: Portfolio.
Carter, C., Clegg, S. R., Kornberger, M. (2013). *A Very Short, Fairly Interesting and Reasonably Cheap Book about Studying Strategy*. London: Sage Publications.
Castells, M. (2001). *The Rise of the Network Society, the Information Age: Economy, Society and Culture*, Vol. I, 2nd edn. Oxford, UK: Blackwell.
Castells, M. (2009). *Communication Power*. Oxford/New York: Oxford University Press.
Centeno, M. A., Nag, M., Patterson, T. S., Shaver, A., Windawi, J. A. (2015). The emergence of global systemic risk. *Annual Review of Sociology*. 41. 65–85.

Chaffee, E. (1985). Three models of strategy. *Academy of Management Review*. 10 (1). 89–98.
Clark, T., Kipping, M. (eds) (2012). *The Oxford Handbook of Management Consulting*. Oxford, UK: Oxford University Press.
Cochet, Y. (2009). *Anti manuel d'écologie*. Paris: Editions Bréal.
Coe, N. M., Yeung, H. W. (2019). Global production networks: Mapping recent conceptual developments. *Journal of Economic Geography*. 19. 775–801.
Coglienese, C. (ed) (2012). *Regulatory Breakdown: The Crisis of Confidence in US Regulations*. Philadelphia, PA: University of Pennsylvania Press.
Deacon, T. (2003b). The hierarchical logic of emergence: Untangling the interdependence of evolution and self-organization. In Weber, F., Depew, B. (eds). *Evolution and Learning*. Cambridge, MA: MIT Press.
Debourdeau, A. (2013). *Les grands textes fondateurs de l'écologie*. Paris: Flammarion.
Dekker, S. (2002). *The Field Guide to Human Error Investigation*. Aldershot, UK: Ashgate.
Dekker, S. (2011). *Drift into failure. From Hunting Broken Components to Understanding Complex Systems*. Aldershot, UK: Ashgate.
Dekker, S. (2018). *The Safety Anarchist. Relying on Human Expertise and Innovation, Reducing Bureaucracy and Compliance*. London: Routldege.
Dekker, S. (2014). The bureaucratization of safety. *Safety Science*. 70. 348–357.
Diamond, J. (1997). *Collapse: How Societies Choose to Fail or Succeed*. New York: Penguin Books.
Diamond, S. (1986a). NASA Cut or Delayed Safety Spending. *The New York Times*. April 24.
Diamond, S. (1986b). NASA Wasted Billions, Federal Audits Disclose. *The New York Times*. April 23.
Diamond, J. (2005). *Collapse. How Societies Choose to Fail or Succeed*. New York: Penguin Books.
Dicken, P. (2015). *Global Shift: Mapping the Changing Contours of the World Economy*, 7th edn. London: Sage Publications.
Downer, J. (2011). '737-cabriolet': The limit of knowledge and the sociology of inevitable failure. *American Journal of Sociology*. 117 (3). 725–762.
Downer, J. (2014). Disowning Fukushima: Managing the credibility of nuclear reliability assessment in the wake of disaster. *Regulation and Governance*. 8. 287–309.
Doz, Y., Kosonen, M. (2010). Embedding strategic agility: A leadership agenda for accelerating business model renewal. *Long Range Planning*. 43 (2/3). 370–382.
Drahos, P., Braithwaite, J. (2001). The globalisation of regulation. *Journal of Political Philosophy*, 9. 103–128.
Dufoix, S. (2013). Les naissances académiques du global. In Dufoix, S., Caillé, A. (eds). *Le tournant global des sciences sociales*. Paris: La Découverte. The social science global turn.
Dumez, S. (2014). La compréhension de l'entreprise entre description, théorie et norme. In Segrestin, B., Roger, B., Vernac, S. (eds). *L'entreprise: Point aveugle du savoir*. Auxerre: Editions Sciences Humaines.
Dupuy, F. (2011). *Lost in Management*. Paris: Le Seuil.
Edmondson, A. (2019). *The Fearless Organization: Creating Psychological Safety in the Workplace for Learning, Innovation, and Growth*. Hoboken, NJ: John Wiley & Sons.
Edmondson, A. C., Verdin, P. (2018). The strategic imperative of psychological safety and organizational error management. In Hagen, J. U. (ed). *How Could This Happen? Managing Errors in Organizations*. London: Palgrave Macmillan.

Ehrlich, P. R., Ehrlich, A. H. (2013). Can collapse of global civilization be avoided? *Proceedings of the Royal Society*. 280. doi:10.1098/rspb.2012.2845.

Eisenhardt, K. M. (2002). Has strategy changed? Mit Sloan. *Management Review*. 43 (2). 88–94.

Eisenhardt, K. M. (2013). Top management teams and the performance of entrepreneurial firms *Small Business Economics*. 40 (4). 805–816.

Eisenhardt, K. M., Furr, N. R., Bingham, C. B. (2010). Microfoundations of performance: Balancing efficiency and flexibility in dynamic environments. *Organization Science*. 21 (6). 1263–1273.

Ellul, J. (1954). *La technique ou l'enjeu du siècle*. Paris: Armand Colin.

Epstein, G. A. (2009). *Financialization and the World Economy*. Cheltenham, UK and Northampton, MA: Edward Elgar Publishing.

Evan, M. W., Manion, M. (2002). *Minding the machines. Preventing techonological disasters*. Upper Saddle River, NJ: Prentice hall.

Farjoun, M., Starbuck, W. H. (2005), Lessons from the Columbia disaster. In Starbuck, H.W., Farjoun, M. (Eds.), *Organization at the Limit. Lessons from the Columbia Disaster*. Malden, MA: Blackwell Publishing.

Farjoun, M. (2005). History and policy at the space shuttle program. In Starbuck, H.W., Farjoun, M. (Eds.), *Organization at the Limit. Lessons from the Columbia Disaster*. Malden, MA: Blackwell Publishing.

Farjoun, M., Starbuck, W. H. (2007). Organizing at and beyond the limits. *Organization Studies*. 28 (4). 541–566.

Finkelstein, S. (2003). *Why Smart Executives Fail: And What You Can Learn from Their Mistakes*. New York: Portfolio.

Fressoz, J. B. (2012). *L'apocalypse joyeuse: Une histoire des risques industriel*. Paris: Le Seuil.

Frischmann, B., Selinger, E. (2018). *Re-Engineering Humanity*. Cambridge: Cambridge University Press.

Fukuyama, F. (1992). *The End of History and the Last Man*. New York: The Free Press.

Ganascia, J. G. (2017). *Le mythe de la singularité: Faut-il craindre l'intelligence artificielle?* Paris: Le Seuil.

Geiselberger, H. (ed) (2017). *L'âge de la régression*. Paris: Folio Essais.

Gell Mann, M. (1995). *The Quark and the Jaguar: Adventures in the Simple and the Complex*. Manhattan, NY: St. Martin's Pres.

Giddens, A. (1990). *The Consequences of Modernity*. Stanford, CA: Stanford University Press.

Giddens, A. (2000). *Runaway World: How Globalization Is Reshaping Our Lives*. New York: Routledge.

Goldin, I., Kurtana, C. (2016). *Age of Discovery: Navigating the Risks and Rewards of Our New Renaissance*. New York: Saint Martin Press.

Goldin, I., Mariathasan, M. (2015). *The Butterfly Defect: How Globalization Creates Systemic Risks, and What to Do about It*. Princeton, NJ: Princeton University Press.

Gomez, P. Y. (2013). *Le Travail Invisible*. Paris: François Bourin Editeur.

Gomez, P. Y. (2018). *La gouvernance d'entreprise*. Paris: Que-sais-je?

Gouldner, A. (1954). *Patterns of Industrial Bureaucracy*. New York: The Free Press.

Grabowsky, M., Roberts, K. H. (2016). Reliability seeking virtual organizations: Challenges for high reliability organizations and resilience engineering. *Safety Science*. 67. 15–20.

Grabowsky, M., Roberts, R. H. (1997). Risk mitigation in large-scale systems: Lessons from high reliability organizations. *California Management Review*. 39 (4). 152–161.
Grant, R. M. (2003). Planning in a turbulent environment: Evidence from the oil majors. *Strategic Management Journal*. 24 (6). 491–417.
Gras, A. (1994). *Grandeur et dépendance*. Paris: Presses Universitaires de France.
Gras, A. (1997). *Les Macrosystèmes techniques*. Paris: Presses Universitaires de France.
Grienvald, J. (2012). Le contexte d'anthropocène et son contexte historique et scientifique. Séminaire du 11 mai 2012. Momentum. Presses Universitaires de France.
Griffin, M., Cordery, J., Soo, C. (2015). Dynamic safety capability: How organizations proactively change core safety systems. *Organizational Psychological Review*. 1 (3). 1–25.
Guénolé, T. (2016). *La mondialisation malheureuse*, 1st edn. Paris: First.
Guthey, E., Clarke, T., Jackson, B. (2009). *Demystifying Business Celebrities*. London and New York: Routledge.
Haavik, T. K. (2016). Remoteness and sensework in harsh environments. *Safety Science*. 95. 150–158.
Hannerz, U. (2003). *Macro-scenarios: Anthropology and the debate over* contemporary and future worlds. *Social Anthropology*. 11 (2). 169–187.
Hamilton, C. (2015). The Risks of Climate Engineering. *The New York Times*. 12th February.
Hamilton, C., Gemenne, F., Bonneuil, C. (2015). *The Anthropocene and the Global Environmental Crisis: Rethinking Modernity in a New Epoch*. Routledge Environmental Humanities series. London and New York: Routledge.
Hamilton, S., Micklewaith, A. (2006). *Greed and Corporate Failure: The Lessons from Recent Disasters*. Basingstoke: Palgrave Macmillan.
Hatch, M.J., Cunliffe, A.L. (2013) *Organization Theory Modern, Symbolic, and Postmodern Perspectives*. 3rd Edition. New York: Oxford University Press.
Hayes, J., Hopkins, A. (2014). *Nightmare Pipeline Failures*. Sydney: CCH.
Hayes, J. 2019. Investigating accidents: the case for disaster case studies in safety science. In Le Coze, JC. (ed) 2019. Safety Science Research. *Evolution, Challenges and New Directions*. Boca Raton, FL: CRC Press, Taylor & Francis Group.
Hay, C. (2006). Globalization and public policy. In Moran, M., Rein, M., Goodin, R. E. (eds). *The Oxford Handbook of Public Policy*. Oxford, UK: Oxford University Press.
Heylinghen, F., Cilliers, P., Gershenson, C. (2007). Complexity and philosophy. In Bogg, J., Geyer, R. (eds). *Complexity, Science and Society*. Oxford, UK: Radcliff.
Hibou, B. (2012). *La bureaucratisation du monde à l'ère néolibérale*. Paris: La découverte.
Hinning, C. R., Greenwood, R. (2002). Disconnects and consequences in organization theory? *Administrative Science Quarterly*. 47. 411–421.
Hirschhorn, L. (1985). On technological catastrophe. *Science*. 228. 846–847.
Hodgkinson, G. P., Starbuck, W. H. (2012). *The Oxford Handbook of Organizational Decision Making*. New York: Oxford University Press.
Hollnagel, E., Woods, D., Leveson, N. (2006). *Resilience Engineering: Concepts and Precepts*. Farnham: Ashgate.
Hollnagel, E. (2014). *Safety-I and Safety-II: The Past and Future of Safety Management. Farnham, UK*: Ashgate.
Hopkins, A. (1978a). The uses of law to sociology. *Australian and New Zealand Journal of Sociology*. 14 (3). 266–273.

Hopkins, A. (1978b). The anatomy of corporate crime. In Wilson, P., Braithwaite, J. (eds). *The Two Faces of Deviance*. Brisbane: University of Queensland Press. 214–231.

Hopkins, A. (1981). Crime without punishment: The Appin mine disaster. *Australian Quarterly*. 53 (4). 455–466.

Hopkins, A. (1984). Blood money? The effect of bonus pay on safety in coal mines, A.N.Z. *Journal of Sociology*. 20 (1). 23–46.

Hopkins, A. (1999a). *Managing Major Hazards: The Lessons of the Moura Mine Disaster*. Sydney: Allen & Unwin.

Hopkins, A. (1999b). The limits of normal accident theory. *Safety Science*. 32. 93–102.

Hopkins, A. (2000). *Lessons from Longford: The ESSO Gas Plant Explosion*. Sydney: CCH Press.

Hopkins, A. (2001). Was Three Mile Island a normal accident? *Journal of Contingencies and Crisis Management*. 9 (2). 65–72.

Hopkins, A. (2005). *Safety, Culture and Risk: The Organizational Causes of Accidents*. Sydney: CCH Press.

Hopkins, A. (2006). Studying organisational cultures and their effects on safety. *Safety Science*. 44. 875–889.

Hopkins, A. (2007a). *Lessons from Gretley: Mindful Leadership and the Law*. Sydney: CCH Press.

Hopkins, A. (2007b). Beyond compliance monitoring: New strategies for safety regulators. *Law and Policy*. 29 (2). 210–225.

Hopkins, A. (2008). *Failure to Learn: The BP Texas City Refinery Disaster*. Sydney, NSW: CCH Press.

Hopkins, A. (2009). *Learning from High Reliability Organisations*. Sydney: CCH Press.

Hopkins, A. (2011). Management walk-arounds: Lessons from the Gulf of Mexico oil well blowout. *Safety Science*. 49 (10 December). 1421–1425.

Hopkins, A. (2012). *Disastrous Decisions: The Human and Organisational Causes of the Gulf of Mexico Blowout*. Sydney: CCH Press.

Hopkins, A. (2016). *Quiet Outrage: The Way of a Sociologist*. Sidney: CCH Press.

Hopkins, A. (2019). *Organising for Safety. How Structure Creates Culture*. Sydney: CCH Press.

Hopkins, A., Palser, J. (1987). The cause of coal mine accidents. *Industrial Relations Journal*. 18 (1 Spring). 26–39.

Hopkins, A., Parnell, N. (1984). Why coal mine safety regulations in Australia are not enforced. *International Journal of the Sociology of Law*. 12. 179–194.

Hopkins, A., Maslen, S. (2015). *Risky Rewards: How Company Bonuses Affect Safety*. London: Ashgate.

Houdart, S., Thiery, O. (ed) (2011). *Humains, non-humains: Comment repeupler les sciences sociales*. Paris: La Découverte.

Hughes, J. J. (2008). Millenial tendencies in responses to apocalyptic threats. In Bostrom, N., Ćirković, M. M. (eds). *Global Catastrophic Risks*. Oxford, UK: Oxford University Press.

Hughes, T. (1989). *American Genesis: A Century of Invention and Technological Enthusiasm, 1870–1970*. New York: Viking.

Hughes, T. (1997). *Rescuing Prometheus*. New York: Pantheon.

Huntington, S. (1996). *The Clash of Civilization and the New World Order*. New York: Simon Schuster.

Jackall, R. (2010). *Moral Mazes: The World of Corporate Managers*, Updated Edition. New York: Oxford University Press.

Jarrige, F. (2016). *Technocritique*. Paris: La découverte.
Jensen, P. (2018). *Pourquoi la société ne se laisse pas mettre en équation?* Paris: Le Seuil.
Joerges, B. (1988). Large technical systems: Concepts and issues. In Mayntz, R., Hughes, T. P. (eds). *The Development of Large Technical Systems*. Frankfurt am Main: Campus Verlag.
Jonas, H. (1984). *The Imperative of Responsibility: In Search of Ethics for the Technological Age*. Chicago: University of Chicago Press.
Jonhson, C.W. (2014). Economic recession and a crisis of regulation in safety-critical industries. *Safety Science*. 68. 153–160.
Johnson, G. (1987). *Strategic Change and the Management Process*. Basil: Blackwell.
Johnson, G., Whittington, R., Scholes, K. (2013). *Exploring Strategy: Texts and Cases*. Harlow: Pearson.
Johnson, N. (2009). *Simply Complexity: A Clear Guide to Complexity Theory*. London: Oneworld Publications.
Kaiser, R. B., Hogan, R., Craig, S. B. (2008). Leadership and the fate of organization. *American Psychologist*. 63 (2). 96–110.
Kay, J. (2015). *Other' People Money: Master of the Universe or Servants of the People?* London: Profile Books.
Kerdellant, C. (2016). *Ils se croyaient les meilleurs: histoire des grandes erreurs de management*. Paris: Denoël.
Kristensen, P. H., Zeitlin, J. (2005). Local players. In *Global Games: The Strategic Constitution of a Multinational Corporation*. Oxford, UK: Oxford University Press.
Kunda, G., Ailon-Souday, G. (2006). Managers, markets and ideologies: Design and devotion revisited. In Ackroyd, S., Batt, R., Thompson, P., Tolbert, P. S. (eds). *The Oxford Handbook of Work and Organization*. New York: Oxford University Press.
La Porte T. R. (1982). On the design and management of nearly error-free organizational control systems. In Sills, D. L., Wolf, C. P., Shelanski, V. B. (eds). *Accident at Three Mile Island: The Human Dimension*. Boulder, CO: Westview Press.
La Porte, T. R. (1994). A strawman speaks up: Comments on the limits of safety. *Journal of Contingencies and Crisis Management*. 2 (4). 207–211.
La Porte, T. R. (ed) (1991). *Social Responses to Large Technical Systems: Control or Anticipation*. Dordrecht: Kluwer.
La Porte, T. R., Consolini, P. (1991). Working in theory but not in practice: Theoretical challenges in high reliability organizations. *Journal of Public Administration Research and Theory*. 1. 19–47.
La Porte, T. R., Rochlin, G. (1994). A rejoinder to Perrow. *Journal of Contingencies and Crisis Management*. 2 (4). 221–227.
Latour, B. (1991). *Nous n'avons jamais été modernes: Essai d'anthropologie symétrique*. Paris: La Découverte.
Le Coze, J. C. (2010). Accident in a French dynamite factory: An example of an organisational investigation. *Safety Science*. 48. 80–90.
Le Coze, J. C. (2012). A study about changes and their impact on industrial safety. *Safety Science Monitor*. 15. 1–17.
Le Coze, J. C. (2015). Reflecting on Jens Rasmussen's legacy: A strong program for a hard problem. *Safety Science*. 71. 123–141.
Le Coze, J. C. (2016). *Trente ans d'accidents: Le nouveau visage des risques sociotechnologiques*. Toulouse: Octarès.

Le Coze, J. C. (2018). Managing the unexpected. In Möller, N., Hansson, S. O., Holmberg, J. E., Rollenhagen, C. (eds). *Handbook of Safety Principle*. London: Wiley. 747–776.

Le Coze, J. C. (ed). (2019a). *Safety Science Research: Evolution, Challenges and New Directions*. Boca Raton, FL: CRC Press, Taylor & Francis Group.

Le Coze, J. C. (2019b). How safety culture can help us think. *Safety Science*. 118. 221–229.

Le Coze, J. C. (2019c). Visualising safety. In Le Coze, J. C. (ed). *Safety Science Research: Evolution, Challenges and New Directions*. Boca Raton, FL: CRC Press, Taylor & Francis Group.

Le Coze, J. C. (2019d). Vive la diversité! High Reliability Organisation (HRO) and Resilience Engineering (RE). *Safety Science*. 117. 469–478.

Le Coze, J. C. (2020). Hopkins' view of structure and culture (on step closer to strategy). *Safety Science*. 122. 104541.

Le Roux, T. (2011). Accidents industriels et régulation des risques: l'explosion de la poudrerie de Grenelle en 1794. *Revue d'histoire moderne et contemporaine*. 58 (3). 34–62.

Lawrence, P. R., Lorsch, J. W. (1967). Differentiation and integration in complex organizations. *Administrative Science Quarterly*. 12. 1–47.

Lewin, R. (1992). *Complexity: Life at the Edge of Chaos*. Chicago: University of Chicago Press.

Lupton, D. (2016). Digital risk society. In Burgess, A., Alemmano, A., Zin, O. (eds). *Routledge Handbook of Risk Studies*. London: Routledge.

Lustgarten, A. (2012). *Run to Failure: BP and the Making of the Deepwater Horizon Disaster*. New York: W.W. Norton & Co.

Maslen, S., Hopkins, A. (2014). Do incentives work? A qualitative study of managers' motivations in hazardous industries. *Safety Science*. 70. 419–428.

Mayer, P., 2003. *Challenger: Les ratages de la décision*. Paris: Presses Universitaires de France.

Mayntz, R., Hughes, T. (eds) (1988). *The Development of Large Technical Systems*. Frankfurt am Main: Campus Verlag.

Meyer, A. D., Starbuck, W. H. (1993). Interactions between ideologies and politics in strategy formation. In Roberts, K. (ed). *New Challenges in Understanding Organisations*. New York: Mc Millan.

Milanovic, B. (2019). Toute unanimité concernant les effets de la mondialisation est impossible. Le Monde. 6 février 2019.

Milch, V., Lauman, K. (2016). Interorganizational complexity and organizational accident risk: A literature review. *Safety Science*. 82. 9–17.

Mills, R. W., Koliba, C. J. (2015). The challenge of accountability in complex regulatory networks: The case of the deepwater horizon oil spill. *Regulation and Governance*. 9. 77–91.

Mintzberg, H. (1973). Strategy making in three modes. *California Management Review*. 16 (2). 4453.

Mintzberg, H., Lampel, J., Ahlstrand, B. (2009). *Strategy Safari: The Complete Guide through the Wilds of Strategic Management*. Philadelphia, PA: Trans-Atlantic Publications.

Mitchell, M. (2009). *Complexity. A Guided Tour*. Oxford, UK: Oxford University Press.

Morgan, G. (2006). Understanding multinational corporations. In Ackroyd, S., Batt, R., Thompson, P., Tolbert, P. S. (eds). *The Oxford Handbook of Work and Organization*. New York: Oxford University Press.

Morgan, G., Whitley, R. (2014). *Capitalisms and Capitalism in the Twenty-First Century*. Oxford, UK: Oxford University Press.
Morin, E. (1977). *La méthode: La nature de la nature*. Paris: Le Seuil.
Morin, E. (2007). Restricted and generalised complexity. In Gershenson, C., Aerts, D., Edmonds, B. (eds). *Worldviews, Science and Us: Philosophy and Complexity*. Singapore: World Scientific.
Norberg, J. (2018). *Non, ce n'était pas mieux avant: 10 bonnes raisons d'avoir confiance en l'avenir*. Paris: Pocket.
Organization for Economic Development and Cooperation. (2003). *Emerging Systemic Risks: Final Report to the OECD Future Project*. Paris: OECD.
Parker, C. (2002). *The Open Corporation: Effective Self-Regulation and Democracy*. Cambridge, UK: CUP.
Perrow, C. (1967). A Framework for the comparative analysis of organizations. *American Sociological Review*. 32. 194–208.
Perrow, C. (1970a). *The Radical Attack on Business: A Critical Review*. San Diego, CA: Harcourt Brace Jovanovich.
Perrow, C. (1970b). *Organizational Analysis: A Sociological View*. London: Tavistock Publications.
Perrow, C. (1973). The short and glorious history of organizational theory. *Organizational Dynamics*. 2 (1). 2–15.
Perrow, C. (1982). The president's commission and the normal accident. In Sills, D. L., Wolf, C. P., Shelanski, V. B. (eds). *Accident at Three Mile Island: The Human Dimension*. Boulder, CO: Westview Press.
Perrow, C. (1983). The organizational context of human factors engineering. *Administrative Science Quarterly*. 28. 521–541.
Perrow, C. (1984). *Normal Accident: Living in High Risk Technology*. New York: Basic Books.
Perrow, C. (1986a). *Complex Organizations: A Critical Essay*, 3rd edn. New York: McGraw Hill.
Perrow, C. (1986b). The Habit of Courting Disaster. *The Nation*. October 1986.
Perrow, C. (1991). A society of organizations. *Theory and Society*. 20. 725–762.
Perrow, C. (1994a). Accidents in high-risk systems. *Technology Studies*. 1. 1–20.
Perrow, C. (1994b). The limits of safety: The enhancement of a theory of accidents. *Journal of Contingencies and Crisis Management*. 2 (4). 212–220.
Perrow, C. (1999). *Normal Accident: Living in High-Risk Technology*, 2nd edn. Princeton, NJ: Princeton University Press.
Perrow, C. (2002). *Organising America: Wealth, Power and the Origins of Corporate Capitalism*. Princeton, NJ: Princeton University Press.
Perrow, C. (2011a). Fukushima, risk, and probability: Expect the unexpected. *Bulletin of the Atomic Scientist*. 67 (6). 44–52.
Perrow, C. (2011b). *The Next Catastrophe: Reducing Our Vulnerabilities to Natural, Industrial and Terrorist Disasters*. Princeton, NJ: Princeton University.
Perrow, C. (2012). Getting to catastrophe. Concentrations, complexity and coupling. The Montreal Review. December. http://www.themontrealreview.com/2009/Normal-Accidents-Living-with-High-Risk-Technologies.php
Perrow, C. (2015). Cracks in the "Regulatory State". *Social Currents*. 2 (3). 203–212.
Pfeffer, J., Sutton, R. (2006). *Management Half-Truths and Nonsense: How to Practice Evidence-Based Management*. Cambridge, MA: Harvard Business Press.

Pidgeon, N. (2019). Observing the English weather: A personal journey from safety I to IV. In Le Coze, J. C. (ed). *Safety Science Research: Evolution, Challenges and New Directions*. Boca Raton, FL: CRC Press, Taylor & Francis Group.

Pinch, T. (1991). How do we treat technical uncertainty in technical failure? The case of the space shuttle Challenger. In La Porte, T. (ed). *Social Responses to Large Technical Systems*, Nato Science Series D. Berlin: Springer. 58. 143–158.

Ponte, S., Gibbon, P. (2005). Quality standards, conventions and the governance of global value chains. *Economy and Society*. 34 (1). 1–31.

Power, M. (1997). *The audit society: rituals of verification*. Oxford, UK: Oxford University Press.

Power, M. (2007). *Organized Uncertainty: Designing a World of Risk Management*. Oxford, UK: Oxford University Press.

Pritchard, S. B. (2012). An envirotechnical disaster: Nature, technology, and politics at Fukushima. *Environmental History*. 17. 219–243.

Quinlan, M. (2015). *Ten Pathways to Disasters: Learning from Fatal Incidents in Mines and Other High Hazard Workplaces*. Annandale: The Federation Press.

Quinlan, M., Hampson, I., Gregson, S. (2013). Outsourcing and offshoring aircraft maintenance in the US: Implications for safety. *Safety Science*. 57. 283–292.

Rasmussen, J. (1990). Human error and the problem of causality in analysis of accidents. *Philosophical Transactions of the Royal Society B*. 327. 449–462.

Rasmussen, J. (1997). Risk management in a dynamic society: A modelling problem. *Safety Science*. 27 (2/3). 183–213.

Reason, J. (1990). *Human Error*. New York: Cambridge University Press.

Reason, J. (1997). *Managing the Risk of Organisational Accidents*. Farnham: Ashgate.

Reason, J. (2005). Preface. In Hopkins, A. (ed). *Safety, Culture and Risk: The Organizational Causes of Accidents*. Sydney: CCH Press.

Reason, J., Mycielska, K. (1982). *Absent-Minded? The Psychology of Mental Lapses and Everyday Errors*. Englewood Cliffs: Prentice Hall.

Rehgezza-Fit, M. (2016). *Des hommes et des risques: Menaces locales, menaces globales*. Paris: La documentation française.

Renn, O. (2016). Systemic risks: The new kid on the block. *Environment: Science and Policy for Sustainable Development*. 58 (2). 26–36.

Roberts, K. H. (1989). New challenges in organisational research: High reliability organizations. *Industrial Crisis Quaterly*. 3. 111–125.

Rochlin, I. G. (1993). Defining 'high reliability' organizations in practice: A taxonomic prologue. In Roberts, K. (Ed.), *New Challenges to Understanding Organizations*. New York: Macmillan Publishing Company.

Rochlin, G. I., La Porte, T. R., Roberts, K. H. (1987). The self-designing high-reliability organization: Aircraft carrier flight operations at sea. *Naval War College Review*. 40 (4). 76–90.

Rodrik, D. (2011). *The Globalization Paradox: Why Global Markets, States and Democracy Can't Coexist*. Oxford, UK: Oxford University Press.

Roe, E., Schulman, P. (2008). *High Reliability Management*. Stanford, CA: Stanford University Press.

Roe, E., Schulman, P. (2016). *Reliability and Risk: The Challenge of Managing Interconnected Infrastructures*. Stanford, CA: Stanford Business Book.

Rosenau, J. N. (2003). Governance in a new global order. In Held, D., McGrew, A. (eds). *The Global Transformations Reader: An Introduction to the Globalization Debate*. Cambridge: Polity Press.

Rouleau, L. (2013). Strategy-as-practice research at a crossroads. *M@n@gement*. 5 (16). 238–252.

Ruet, J. (2016). *Des capitalismes non alignés: les pays émergents, ou la nouvelle relation industrielle du monde*. Paris: Raisons d'Agir.

Sagan, S. D. (1993). *The Limits of Safety: Organizations, Accidents, and Nuclear Weapons*. Princeton, NJ: Princeton University Press.

Sagan, S. D. (1994). Towards a political theory of organizational reliability. *Journal of Contingencies and Crisis Management*. 2 (4). 228–240.

Sassen, S. (2006). *Territory, Authority, Rights: From Medieval to Global Assemblages*. Princeton, NJ: Princeton University Press.

Sassen, S. (2007). *A Sociology of Globalization*. New York: W. W. Norton & Co.

Sassen, S. (2014). *Expulsions: Brutality and Complexity in the Global Economy*. Cambridge, MA: Belknap Press.

Sayles, S., Smith, C. (2006). Rogue executive. In *The Rise of the Rogue Executive: How Good Companies Go Bad and How to Stop the Destruction*. New York: Pearson, Prentice Hall.

Selznick, P. (1949). *TVA and the Grass Roots; A Study in the Sociology of Formal Organization*. Berkeley and Los Angeles: University of California Press.

Schein, E. H. (1992). *Organizational Culture and Leadership*. San Francisco, CA: Jossey-Bass Publishers.

Schulman, P., Roe, E. (2018). Extending reliability analysis across organizations, time, and scope. In Ramanujam, R., Roberts, K. (eds). *Organizing for Reliability: A Guide for Research and Practice*. Stanford CA: Stanford University Press. 194–214.

Scott, C. (2004). Regulation in the age of governance: The rise of the post regulatory state. In Jordana, J., Levi-Faur, D. (eds). *The Politics of Regulation: Institutions and Regulatory Reforms for the Age of Governance*. Cheltenham, UK: Edward Edgar.

Scott, W. R. (1995). *Institutions and Organizations: Ideas, Interests and Identities*. Thousand Oaks, CA: Sage Publications.

Scott, W. R. (2003). *Organization, Rational, Natural and Open Systems*, 5th edn. Upper Saddle River, NJ: Prentice Hall.

Sennett, R. (2006). *The Culture of the New Capitalism*. New Havens, CO and London. New University Press.

Serres, M. (2015). *Pantopie ou le monde selon Michel Serres: de Hermès à petite poucette (entretiens par Martin Legros et Sven Ortoli)*. Paris: Le Pommier.

Servigne, P., Stevens, R. (2015). *Comment tout peut s'effondrer: Petit manuel de collapsologie à l'usage des générations présentes*. Paris: Le Seuil.

Shimizu, K., Hitt, M. (2011). Errors at the top of the hierarchy. In Hofmann, D. A., Frese, M. (eds). *Errors in Organizations*. New York: Routledge.

Snook, S. A. (2000). *Friendly Fire, the Accidental Shoot Down of US Black Hawks over Northern Irak*. Princeton, NJ: Princeton University Press.

Starbuck, H. W., Farjoun, M. (eds) (2005). *Organization at the Limit: Lessons from the Columbia Disaster*. Hoboken, NJ: Blackwell Publishing.

Starbuck, W. H., Milliken, F. J. (1988a). Challenger: Changing the odds until something breaks. *Journal of Management Studies*. 25. 319–340.

Starbuck, H. W., Milliken, F. J. (1988b). Executives' perceptual filters: What they notice and how they make sense. In Hambrick, D. C. (ed). *The Executive Effect: Concepts and Methods for Studying Top Managers*. Greenwich, CT: JAI Press. 35–65.

Starbuck, W. H. (2009). Cognitive reactions to rare events: Perceptions, uncertainty, and learning. *Organization Science*. 20. 925–937.

Starbuck, W. H., Greve, A., Hedberg, B. (1978). Responding to crises. *Journal of Business Administration*. 9 (2). 111–137.
Staw, B. M. (1997). The escalation of commitment: An update and appraisal. In Shapira, Z. (ed). *Organizational Decision Making*. Cambridge: Cambridge University Press. 156–189.
Steffen, W., Grinevald, C. P., McNeill, J. (2011). The anthropocene: Conceptual and historical perspectives. *Philosophical Transactions of the Royal Society A*. 369. 842–867.
Stiglitz, J. E. (2003). *The Roaring Nineties: A New History of the World's Most Prosperous Decade*. New York: W.W. Norton & Company.
Stiglitz, J. E. (2010). *Freefall: America, Free Markets, and the Sinking of the World Economy*. New York: W.W. Norton & Company.
Sturdy, A. (2011). Consultancy's consequences? A critical assessment of management consultancy's impact on management. *Bristish Journal of Management*. 22. 517–530.
Susuki, A., Nakai, A. (2012). *Fuskuhima: Chronique d'un désastre*. NHK International Inc.
Sutcliffe, K. (2018). Mindful organizing. In Ramanujam, R., Roberts, K. (eds). *Organizing for Reliability: A Guide for Research and Practice*. Stanford, CA: Stanford University Press. 194–214, 61–89.
Teece, D. J. (2007). Explicating dynamic capabilities: The nature and microfoundations of (sustainable) enterprise performance. *Strategic Management Journal*. 28 (13). 1319–1350.
Teece, D., Peteraf, M., Leih, S. (2016). Dynamic capabilities and organizational agility: Risk, uncertainty, and strategy in the innovation economy. *California Management Review*. 58 (4) 13–35.
Therborn, G. (2011). *The World: A Beginner's Guide*. Cambridge: Polity Press.
Turner, B. (1992). The sociology of safety. In Blockley, D. I. (ed). *Engineering Safety*. Maidenhead: Mc Graw-Hill.
Turner, B. A. (1978). *Man-Made Disaster: The Failure of Foresight*. London: Wykeham Science Press.
Urry, J. (2005). The complexity turn. *Special Issue of Theory Culture and Society*. 22 (1). 1–14.
Vaara, E., Whittington, R. (2012). Strategy-as-practice: Taking social practices seriously. *The Academy of Management Annals*. 6 (1). 285–336.
van der Zwan, N. (2014). Making sense of globalization. *Socio-Economic Review*. 12. 99–129.
Vaughan, D. (1990). Autonomy, interdependence, and social control: NASA and the Space Shuttle Challenger. *Administrative Science Quarterly*. 35. 225–257.
Vaughan, D. (1996). *The Challenger Launch Decision: Risky Technology, Culture and Deviance at NASA*. Chicago, IL: University of Chicago Press.
Vaughan, D. (1997). The trickle-down effect: Policy decision, risky work, and the challenger tragedy. *California Management Review*. 39 (2). 80–102.
Vaughan, D. (1999). The dark side of organizations: Mistake, misconduct, and disaster. *Annual Review of Sociology*. 25. 271–305.
Vaughan, D. (2005). System effects: On slippery slopes, repeating negative patterns, and learning from mistakes? In Starbuck, H. W., Farjoun, M. (ed). *Organization at the Limit: Lessons from the Columbia Disaster*. Hoboken, NJ: Blackwell Publishing.
Vaughan, D. (2011). *Theorizing the Dark Side: Special Issue on the Dark Side*. Organization Studies.

Veltz, P. (2008). *Le nouveau monde industriel: Edition revue et augmentée*. Paris: Le débat - Gallimard.
Vogel, D. (2008). Private global business regulation. *Annual Review of Political Science*. 11. 261–282.
Waldrop, M. (1992). *Complexity: The Emerging Science at the Edge of Order and Chaos*. New York: Simon & Schuster.
Wallerstein, I. (2004). *Introduction to World System Analysis*. Durham, NC: Duke University Press.
Waterson, P. E., Lemalu Kolose, S. (2010). Exploring the social and organisational aspects of human factors integration: A framework and case study. *Safety Science*. 48 (4). 482–490.
Wears, R. L., Hunte, G. S. (2014). Seeing patient safety' like a state. *Safety Science*. 67. 50–57.
Weick, K. (1995). *Sensemaking in Organizations*. London: Sage Publications.
Weick, K., Sutcliff, K. M., Obstfeld, D. (1999). Organising for high reliability: Processes of collective mindfullness. *Research in Organisational Behavior*. 21. 81–123.
Woods, D., Johannesen, L., Cook, R. I., Sarter, N. (1994). *Behind Human Error: Cognitive System, Commuters and Hindsight*. Wright-Patterson AFB, OH: Crew System Ergonomics.
Wynne, B. (1988). Unruly technology: Practical rules, impractical discourses and public understanding. *Social Studies of Science*. 18. 147–167.
Yeung, H. W., Coe, N. M. (2014). Towards a dynamic theory of global production networks. *Economic Geography*. 91 (1). 29–58.
Yeung, K. (2017). Algorithmic regulation: A critical interrogation. *Regulation and Governance*. 12. 505–523.
Zakaria, F. (2012). *The Post-American World: Release 2.0*. London: Norton & Company.

Index

A

AcciMap layout, 36, 37
Adaptive strategy, 75–76
Adequate strategy, 91
AI, *see* Artificial intelligence (AI)
Analysing strategic failures, 79–81
Anthropocene, x, 129–130, 138, 144
Artificial intelligence (AI), 132, 138
Auditing approach, 44–49
"Authoritarian high modernism," 108
Aviation industries, safety in, 25, 26

B

British Petroleum (BP), 32, 68, 69, 150
  globalised trends in, 120–121
  postbureaucratic strategy failure, 119–120
  postregulatory state failures, 119–120
  strategic fiasco of multinational, 115
    Browne's legacy, 116–117
    networked firm, 118–119
Bureaucratic organization, 14
Bureaucratisation, of safety, 107
BUs, *see* Business units (BUs)
Business strategy, 79
Business units (BUs), 90, 92, 93, 95, 112, 116, 117, 119, 120

C

Catastrophic events, 54, 59, 94
Causal circularity
  globalisation
    and anthropocene/ transhumanism, 138
    high-risk systems and, 137–138
Centralised structure, 60, 61
Challenger explosion, 4, 5
Cognitive engineering, 73, 74
Cognitive layer, *NA*, 22–23
Colonisation, 109
Complexity, eco-socio-technological disasters, xii, 142–144
Component accidents, 5, 99; *see also* British Petroleum (BP)
  globalisation, 99, 100, 123
    good/bad, 102
    to high-risk systems, 103–105, 122
    Hopkins' view and, 113–115
    overview of, 100–101
    in safety, 104–111
    from sceptics, 101–102
    uniformity process, 102–103
  to network failure accidents, 121–123
Component failure accidents, 149–150
Conceptualise safety, 83
Contingency theory, 13, 27
Cost-cutting strategy, 91–93
Coupling/interactivity framework, 21, 22, 33

D

Dark side of organization (Vaughan), 7
Decentralised organisations, 119
Decentralising safety, 120
Decision makers, 80
Decision-making process, xii, 41, 54, 62, 80, 84, 95, 96, 108, 112, 149
Decoupling errors, 74
Degree of strategic failure, 81–82
DEPOSE (design, equipment, procedures, operators, supplies and materials, and environment) system, 5, 20, 24, 122
Deregulation context, 117
Digitalisation, 109–110, 120
Digitalised risks, 129
Disaster/safety model, 56

E

Eco-socio-technological systems
  causal circularity, globalisation
    and anthropocene/
      transhumanism, 138
    high-risk systems and, 137–138
  embedded risk categories, complex
    interactions, 136
  Fukushima Daïchi, 141–142
  Normal Accident, 125, 139, 141, 142,
    144, 145
  overview of, 125–126
  scope, scale and time frame, 134
    anthropocene and
      transhumanism, 129–132
    existential risks, 129–132
    globalisation and systemic risks,
      127–129
    high-risk systems, 126–127
    sociotechnological risks, 126–127
    widening and complexifying,
      132–135
Ecosystem accident, 133, 141, 142
"Ellulian" thread, 5
Embedded risk categories
  complex interactions, 136
  globalisation
    and anthropocene/
      transhumanism, 138
    high-risk systems and, 137–138
Envirotechnical disaster, 142, 144
Error-avoiding system, 23, 25, 26
Error-inducing system, 20, 26, 29
Escalation of commitment, 80, 81, 84
Esso's cost-cutting, 39
Existential risks, 130–132
Externalisation, 105–106, 120–121

F

Failsafe system, 141–142
Federal Aviation Authority, 106
Financialisation, 108–109, 111, 120, 121
Flixborough accident, 29
Framing strategy, 78
Front-line actors, 67, 74, 149
Front-line operators, 73, 74, 96
Fukushima Daïchi, 11, 18, 31, 137,
    141–142, 144, 150
Functionalism, society of organisations,
    15–16

G

General complexity, eco-socio-
    technological disasters, xii,
    142–144, 150
Global catastrophic risks, 132
Globalisation, ix, xii, 99, 100, 123, 126,
    127–129, 149–150
  and anthropocene/
    transhumanism, 138
  good/bad, 102
  to high-risk systems, 103, 105, 122
    postbureaucracy, 104
    postregulatory state, 104
  high-risk systems and, 137–138
  Hopkins' view and, 113–115
  overview of, 100–101
  in safety, 104, 105
    digitalisation, 109–110
    externalisation, 105–106
    financialisation, 108–109, 111
    self-regulation, 110–111
    standardisation, 106–108, 111, 114
  from sceptics, 101–102
  uniformity process, 102–103
Global risks, 128

H

Health and safety law, 50
High-reliability organisations (HRO),
    32, 55
  vs. normal accident, 61–65
High-risk systems, x, xi, 2, 3, 7, 8, 23, 25,
    28, 30, 41, 56, 63, 125
  eco-socio-technological systems,
    126–127
  globalisation to, 103–105, 122
  operating landscape of, 105,
    121–123, 150
  safety in, xii, 148–149
  sociological study of, ix
Hindsight bias, 24, 94–96

Hopkins' accident diagram, strategy, 70
Hopkins, Andrew, 5
  AcciMap of Longford disaster, 36
  extended version of *NA*, 7–8
  high-reliability organisation, 55
    vs. normal accident, 61–65
  Longford disaster, 36, 58
  a narrative structure, 43–44
    auditing approach, 44–49
  Normal Accident, second thesis of
    complexity argument, 58–59
    sociology of safety, 59–61
  overview of, 35–36
  refutation of *Normal Accident*, 5–7
  second thesis theory
    normative theory of safety, 51–54
    organisations' environment, 55
    safety culture, 56–57
    structure and goal, 55
    technology, 54–56
    white-collar crime model of accident, 49–51
  storytelling skills, 36, 43
  visualising accidents, 36
    analysis and causation, 36–41
  vs. Perrow's visualisations, 41–43
HRO, *see* High-reliability organisations (HRO)
Human errors, 6, 22, 67, 73–74
  of front-line actors, 67, 149

I

ICT, *see* Information and communication technology (ICT)
Incorrect bypass operation, 38, 41
Industrial revolution, 134
Information and communication technology (ICT), 104, 105, 122
Interpretive strategy, 75–76

L

Large technical systems, 126
Linear strategy, 75–76
Longford disaster, 36, 58
  AcciMap of, 36, 37
  physical accident sequence, 37, 38

M

Machiavelli's view of power, 77
Macondo project, 118
Macrosystem layer, *NA*, 25, 26
Magnitude of impact, Perrow's matrix, 139–140
Marine industries, safety, 25, 26
Market privatisation ideology, 40
Marx, Karl, 15, 18
Mistake, failure and fiasco, in safety strategy, 92–94
Multidimensional approach, *Normal Accident*, 27, 33

N

NA, see Normal Accident (NA)
National Transportation Safety Bureau, 106
"Neo-liberal paradox," 111
Neo-Weberian approach, 21–22
Networked configuration, 90
Network failure accidents, xii, 121–123, 149–150
New reductionism, 96–98
Normal Accident (NA), ix–x, 1, 29–30
  component accidents *vs.*, 5
  eco-socio-technological systems, 125, 139, 141, 142, 144, 145
  extended version of, 7–8
  high-risk systems, 2–3
  Hopkins, Andrew
    complexity argument, 58–59
    refutation of, 5–7
    sociology of safety, 59–61, 148–149
  multidimensional approach, 27
  new version of, 31
  operating landscape, 112–113
  safety and disaster research, 28
  second thesis, 18–21
    Challenger case, 21
    cognitive layer, 22–23
    implications of, 28–31
    macrosystem layer, 25, 26
    neo-Weberian approach, 21–22
    organisational/managerial layer, 23–24
    tension sources, 19–20

Normal Accident (*cont.*)
second thesis of, 148–149
singling out, 2–3
technological determinism, 4–5
vs. Post Normal Accident, 112–113, 147–148
Normative theory of safety, 51–54
Nuclear power plants, ix
"Nuclear winter" threat, 135
Nutshell, *Post Normal Accident* in, xi–xii

O

Obviousness
of strategy
failing executives and corporate malfeasance, 68–71
organisations at and beyond the limits, 71–73
Operating landscape
financialisation, 109
of high-risk systems, 105, 121–123, 150
Normal Accident, 112–113
Organisational/managerial layer, *NA,* 23–24
Organisational structures, 58, 78, 91, 96, 114
Organisational theory, 10, 12, 14, 22, 62, 63
"Organisations at the limits," 24, 71

P

Perrow, Charles, 4
sociology of organisations, 33
context of 1960s, 11–13
goals, 14
limitations, 31–32
related books and studies, 9–11
society of organisations, 15–17
technology (task) and structure, 13–14
Perrow's matrix, 125
eco-socio-technological disasters
complexity, 142–144
Fukushima Daïchi, 141–142
scale of governance
and magnitude of impact, 139–140
Perrow's 2×2 matrix, xi, xii, 126, 139, 146
Petro case, 89–90, 92, 93
Petrochemical industry, 43
Piper Alpha, 45
Pluralism, society of organisations, 15–16
Poor auditing, a narrative structure, 44, 46–48
Postbureaucracy, xi
Postbureaucratic strategy failure, 119–120
Postcolonialism, xi
Posthumanism, xi
Postindustrial society, x
Postmodernity, x
Postnormal science, xi
Postregulatory state failures, 119–120
Practical drift (Snook), 7, 8
Preglobalisation systemic risks, 135
Private organizations, 17, 18
Pyro case, 87–89, 91, 93

R

The Radical Attack on Business, 15–16
Reductionism, 96–98, 143
"Regulatory capture," 32
Retrospective fallacy, 24, 94–96
Risk, 81
homeostasis, 24
"Rogue executive," 82

S

Safety, xi
in aviation industries, 25, 26
globalisation, 104, 105
digitalisation, 109–110
externalisation, 105–106
financialisation, 108–109, 111
self-regulation, 110–111
standardisation, 106–108, 111, 114
in marine industries, 25, 26
as strategy
for businesses, 74–75
executives and top managers interaction, 76–78
hindsight bias, 94–95
illustration, 84, 85

linear, adaptive and interpretive views, 75–76
mistake, failure or fiasco, 92–94
new reductionism, 96–98
petro case, 89–90
pyro case, 87–89
silo case, 86–87
three cases from strategic angle, 90–92
Safety auditing, 45
Safety discourse strategy, 67
Safety management system, 44–47
Scale of governance, Perrow's matrix, 139–140
Score card approach, 45
Second thesis theory
Normal Accident, 18–21
Challenger case, 21
cognitive layer, 22–23
implications of, 28–31
macrosystem layer, 25, 26
neo-Weberian approach, 21–22
organisational/managerial layer, 23–24
tension sources, 19–20
normative theory of safety, 51–54
organisations' environment, 55
safety culture, 56–57
structure and goal, 55
technology, 54–56
Self-regulation, 110–111, 121
Silo case, 86–87, 91
Society of organisations, 9, 10, 15–17
Socioengineering, 46
Sociology, x, 4, 6, 11, 31, 48, 148
Sociotechnological risks, 126–127, 132, 135, 136
Standardisation, 106–108, 111, 114, 121
Strategic breakdowns, 98
Strategic dissonance, 80, 84
Strategic drift, 80, 84
Strategic mistakes, 81–2, 92–94
Strategic orientations, 74, 77, 79, 89–91, 95, 98
Strategy
as core dimension, 97
decision-making, xii, 80, 82, 112, 118, 149
definition of, 82
failures
analysis, 79–81
degree, 81–82
in post normal accident, 82–84
framing, 78
Hopkins' accident diagram, 70
implementation problem, 95–96
obviousness of
failing executives and corporate malfeasance, 68–71
organisations at and beyond the limits, 71–73
in post normal accident, 82–84
safety
for businesses, 74–75
executives and top managers interaction, 76–78
hindsight bias, 94–95
illustration, 84, 85
linear, adaptive and interpretive views, 75–76
mistake, failure or fiasco, 92–94
new reductionism, 96–98
petro case, 89–90
pyro case, 87–89
silo case, 86–87
three cases, 90–92
Sustainable development, 116
System accident, 5, 20, 28
Systemic risks, 127–129, 135, 137

T

Technological determinism, *Normal Accident,* 4–5
Three Miles Island (TMI), 6, 10, 20, 29, 126, 148
Top decision-makers, 59, 65, 67, 74, 90, 98, 99
Transhumanism, 130, 138
"Trickle down effect," 71

V

Visualisations, of Perrow and Hopkins, 41–43
Visualising accidents, 36–41

W

Wallerstein's world systems view, 101
WCC, *see* White-collar crime (WCC)
Weber, Max, 15, 18
White-collar crime (WCC), 35, 43, 49–53, 65
Widespread management system, 121

Printed in the United States
by Baker & Taylor Publisher Services